# OUR WIRED NERVES

# OUR WIRED NERVES

## The Human Nerve Connectome

**DOUGLAS W. ZOCHODNE**

Professor and Director, Division of Neurology,
Department of Medicine, University of Alberta,
Edmonton, Alberta, Canada

ELSEVIER

ACADEMIC PRESS

An imprint of Elsevier

Academic Press is an imprint of Elsevier
125 London Wall, London EC2Y 5AS, United Kingdom
525 B Street, Suite 1650, San Diego, CA 92101, United States
50 Hampshire Street, 5th Floor, Cambridge, MA 02139, United States
The Boulevard, Langford Lane, Kidlington, Oxford OX5 1GB, United Kingdom



**Notices**
Knowledge and best practice in this field are constantly changing. As new research and experience broaden our understanding, changes in research methods, professional practices, or medical treatment may become necessary.

Practitioners and researchers must always rely on their own experience and knowledge in evaluating and using any information, methods, compounds, or experiments described herein. In using such information or methods they should be mindful of their own safety and the safety of others, including parties for whom they have a professional responsibility.

To the fullest extent of the law, neither the Publisher nor the authors, contributors, or editors, assume any liability for any injury and/or damage to persons or property as a matter of products liability, negligence or otherwise, or from any use or operation of any methods, products, instructions, or ideas contained in the material herein.

**Library of Congress Cataloging-in-Publication Data**
A catalog record for this book is available from the Library of Congress

**British Library Cataloguing-in-Publication Data**
A catalogue record for this book is available from the British Library

ISBN 978-0-12-821487-9

For information on all Academic Press publications
visit our website at https://www.elsevier.com/books-and-journals

*Publisher:* Nikki Levy
*Acquisitions Editor:* Joslyn Chaiprasert-Paguio
*Editorial Project Manager:* Tracy I. Tufaga
*Production Project Manager:* Omer Mukthar
*Cover Designer:* Mark Rogers
*Cover Image Artist:* Dr. Marilène Oliver

Typeset by SPi Global, India

# Contents

# Author's biography

**Dr. Douglas W. Zochodne** is a Neurologist and Neuroscientist, Divisional Director of Neurology at the University of Alberta (UofA) and Director of the Neuroscience and Mental Health Institute, UofA. He was trained in Neurology and Neuromuscular Diseases the University of Western Ontario (now Western University) and at Mayo Clinic. Dr. Zochodne's career has included faculty positions at Queen's University, Canada, the University of Calgary and more recently the University of Alberta. He has devoted his career toward understanding the biology and diseases of the peripheral nervous system. Dr. Zochodne's roles have included Editor-in-Chief of the Canadian Journal of Neurological Sciences (1999–2007), and President of the Peripheral Nerve Society (2009–2011). He is a Fellow of the Canadian Academy of Health Sciences, was awarded the Wolfe Prize in Neuropathy research by the American Neurological Association in 2011 and was awarded the Alan J. Gebhart Prize for Excellence in Peripheral Nerve Research, by the Peripheral Nerve Society in 2019. Dr. Zochodne has channeled his research as an investigator of diabetic polyneuropathy and peripheral neuron regeneration. His laboratory has been funded continuously since 1988 and it has published over 285 research papers and chapters including four books.

# Preface

Peripheral neurobiology combines the beauty of exquisite neurosciences and cell biology with a clinical imperative. I have had the privilege of working within this field for over 30 years. I would not characterize my own research findings as leading the field or shaking its foundations, but it has been a labor of love and joy. I hope to highlight a few of the persons who have shaken the field, both in descriptions of the biology through the text and in a dedicated chapter at the end. In this book, my aim is to convey a sense of this field, far too long "below the radar". The term "peripheral neuropathy" is simply not part of the popular lexicon despite television spots that advocate treatment for "diabetic nerve pain".

There are some significant challenges that will require work in moving the field forward and into public awareness. Lack of appreciation of the peripheral connectome does permeate from lacunae in popular thinking into editorial and grant funding decisions. My colleagues and I continue to cringe at comments from Editors-in-Chief and Section Editors suggesting redirection to a "more specialized or suitable journal" despite their previous publications around biology and disease that is logarithms less common or fundamental. Diabetic neuropathy has been a special case in point. The disorder is rarely recognized among the common neurological disorders being studied and is usually dismissed to the endocrine world as a result. However, among diabetic endocrine researchers and clinicians, polyneuropathy is an afterthought during thinking about pancreatic function, vascular disease, retinopathy or nephropathy. Earlier in my career I took it upon myself to check the frequency of publication on the regeneration, biology or disorders of adult peripheral nerve in key leading high impact scientific publications (you know the titles) over the previous ten years. Zero. Few members of their Editorial boards had been close to the field of peripheral neurobiology. The important news is that there have been some improvements in staffing and in the publication of high quality work on Schwann cells and regeneration. My hope is that this would continue to grow. Sometimes researchers "sell" their PNS work by suggesting how helpful it might be for solving a CNS disorder! The final challenge in all neuroscience research is to convince society and governments over the pressing need for unfettered discovery research. I have admitted my biases over the relative importance of discovery and targeted research in the writing of this book. Discovery work yields advances that dwarf those of limited targeted goals. In Canada, Dr. David Naylor and colleagues have extensively documented this shortcoming in a key report for government. However, we worry about its continued traction on decision makers.

The challenge in clinical work has been in conveying a need for research to make a difference in the burdens of patients. As of 2020, the lack of carefully tested therapy in a majority of neuropathies remains the state of affairs. There have been exceptional breakthroughs however, such as new treatments for inherited amyloid neuropathy. We hope these changes will permeate into other difficult neuropathies. Diabetic polyneuropathy, the most common of the acquired neuropathies remains irreversible and without specific treatment. Similar comments apply to the CMTs.

This book is designed to be an accessible overview. The idea is to hint at the complexity of many of the molecules and cells involved but the danger may be in oversimplifying these concepts. There are many molecules involved and to eliminate too many would be remiss. I apologize at the outset for omissions in recognizing work, investigators, molecules or other aspects of a number of the areas covered. It is close to impossible to summarize such a large field with perfect accuracy and attribution. I have not spent significant time covering the surgical approaches to nerve repair, an area of extensive discussion among hand, plastics and neurosurgeons. Tom Brushart's text, "Nerve Repair" is a far better source for this material. The citations here provide a guide to further in depth reading by those interested and prospective students of the area. I hope there will be many.

My final message would be to encourage workers in this field not to give up. Do not abandon peripheral neurobiology for areas that may be less common but appear higher on the radar. Join the Peripheral Nerve Society and be exposed to an amazing cohort of scientists, physicians and others. For trainees, this is an important and very rewarding area of work that speaks to the need for knowledge and treatment. For persons suffering from neuropathy, do not give up hope, there is progress and your advocacy will make an important difference. For governments, funding agencies and leading journals, pay attention! This is important!

**Douglas W. Zochodne**

# Acknowledgments

This book is dedicated to my spouse, Barbara who spent untold hours providing detailed editorial input. Her expertise brought general advice, clarity, brevity and many other qualities to the writing. It is also dedicated to my children, Julia and William who continue to inspire me with their success and friendship.

I am grateful to several people who allowed reproduction of key images especially Drs. Marilène Oliver (cover image), Alain Chédotal, Peter Dyck, Charles Bolton and Kilgore Corporation. Rebecca Long provided an expert review of the text.

Current and previous members of my own laboratory contributed to outstanding research, some of which is highlighted here including Anand Krishnan, Ambika Chandrasekhar, Arul Duraikannu, Masaki Kobayashi, Shane Eaton, Bhagat Singh, Vandana Singh, Chu Cheng, James Kennedy, David McDonald, Trevor Poitras, Jesi Bautista, Aparna Areti, Kaylynn Purdy, Prashanth Komirishetty, Matt Larouche, Kim Christie, Christine Webber, Jose Martinez, Dan Levy, GuiFang Guo, Valentine Brussee, Yuan Yuan Chen, Hong Sun, QG Xu and a number of additional outstanding individuals. Regrets I have not included everyone over a 30 year career with colleagues.

The publication staff at Elsevier were very helpful in planning this book including Joslyn, Tracy and Ashwathi. At the University of Alberta I have had tremendous support for my work from Dr. Barbara Ballermann, the Department Chair of Medicine, the University Hospital Foundation, Ms. Susan Tiller and my Neurology Divisional colleagues. My research has been supported by the Canadian Institutes of Health Research, Diabetes Canada (formerly Canadian Diabetes Association), the Alberta Heritage Foundation for Medical Research, the Juvenile Diabetes Research Foundation, the Alberta Diabetes Institute, Faculty of Medicine and Dentistry, Department of Medicine, and the Division of Neurology all at the University of Alberta. The Neuroscience and Mental Health Institute of the University of Alberta remains my academic home. A special thanks for colleagues of the Peripheral Nerve Society for their support and interactions.

Finally I have had the privilege of seeing and working with many persons dealing with disorders of the PNS over my career. Their forbearance and courage has been inspiring. Examples have included Brian Langton who dealt with the aftermath of severe GBS, members of the Calgary

Neuropathy Association [https://calgaryneuropathy.com/], a patient support group led by Sylvia Donley and founded by Val Shaw-Lewis, and members of the GBS/CIDP Foundation of Canada [https://www.gbscidp.ca/]. The patients have taught me a great deal about the need to do better. I hope that will be the case for all of us in the field.

# Abbreviations

| | |
|---|---|
| **Ach** | acetylcholine |
| **AchR** | acetylcholine receptor |
| **AP** | action potential |
| **APC** | adenomatous polyposis coli |
| **ATF-3** | cyclic AMP-dependent transcription factor |
| **AxD** | axonal degeneration |
| **BDNF** | brain-derived neurotrophic factor |
| **BMP** | bone morphogenetic protein |
| **BOLD** | blood-oxygen-level dependent |
| **CNS** | central nervous system |
| **CNTF** | ciliary neurotrophic factor |
| **CRPS** | chronic regional pain syndrome |
| **CSPGs** | chondroitin sulfate proteoglycans |
| **DADS** | distal acquired demyelinating symmetric polyneuropathy |
| **DCC** | deleted on colorectal carcinoma (protein) |
| **DM** | diabetes mellitus |
| **DPN** | diabetic polyneuropathy |
| **EGF** | epidermal growth factor |
| **EM** | electron microscopy |
| **CGRP** | calcitonin gene-related peptide |
| **CIP** | critical illness polyneuropathy |
| **CMAP** | compound muscle action potential |
| **CMT** | Charcot-Marie-Tooth disease |
| **CPG** | central pattern generator |
| **GABA** | gamma aminobutyric acid |
| **GAD** | glutamic acid decarboxylase |
| **GBS** | Guillain-Barré syndrome |
| **GDNF** | Glial cell line-derived neurotrophic factor |
| **GPCR** | G protein-coupled receptors |
| **GTT** | glucose tolerance test |
| **HGF** | hepatocyte growth factor |
| **IL1,6** | interleukin 1, or 6 |
| **IR** | insulin receptor |
| **HMSN** | hereditary motor and sensory neuropathy (CMT) |
| **IGOS** | International GBS Outcome Study |

| | |
|---|---|
| **MAG** | myelin associated glycoprotein |
| **MCL** | miraculin |
| **MDT** | multidrug therapy |
| **MEPP** | miniature endplate potential |
| **MFMN** | multifocal motor neuropathy |
| **MFS** | Miller-Fisher syndrome |
| **MG** | myasthenia Gravis |
| **MM** | mononeuropathy multiplex |
| **MS** | multiple sclerosis |
| **NAD** | nicotinamide adenine dinucleotide |
| **NCAM** | neural cell adhesion molecule |
| **NE** | norepinephrine |
| **NF-κB** | nuclear factor kappa-light-chain-enhancer of activated B cells |
| **NGF** | nerve growth factor |
| **NMJ** | neuromuscular junction |
| **NMNAT** | nicotinamide mononucleotide adenylyltransferase |
| **OMgp** | oligodendrocyte myelin protein |
| **PDGF** | platelet-derived growth factor |
| **PNS** | peripheral nervous system or Peripheral Nerve Society |
| **PSA** | polysialic acid |
| **PTN** | pleiotrophin |
| **PTEN** | phosphatase and tensin homolog deleted on chromosome ten |
| **Rb1** | retinoblastoma1 |
| **rhNGF** | recombinant human nerve growth factor |
| **RIPK** | receptor-interacting protein kinase |
| **SC** | Schwann cell |
| **Shh** | Sonic Hedgehog protein |
| **SKP** | skin derived precursor cell |
| **SMA** | spinal muscular atrophy |
| **SMN1,2** | survival motor neuron 1,2 (protein) |
| **SNAP** | sensory nerve action potential |
| **SP** | substance P |
| **TN** | trigeminal neuralgia |
| **TRADD** | tumor necrosis factor receptor type 1-associated death domain |
| **TRAF** | TNF receptor-associated factor 1 |
| **TNF** | tumor necrosis factor |
| **TRP** | transient receptor potential (protein) |
| **Trk** | tropomyosin receptor kinase |
| **Unc** | uncoordinated (protein) |
| **WHO** | World Health Organization |

# Introduction

It is possible you have never heard of the peripheral nervous system. Although it is also called the PNS for short, that does not help much. When thinking of the nervous system, there of course is the brain and with it there is thought, control and action. The brain is often described as the central player, the director and master of all human activity. The brain is connected to the spinal cord, allowing commands and inputs to travel outward from the control center. Beyond the spinal cord and the CNS however, how the "house is wired" is often not considered. This is the forgotten part of the nervous system and the topic of this book.

Here we describe the story around "nerves," the neglected but critical links between the brain and the body. By "nerves" the proper scientific terminology is the peripheral nervous system (PNS). The PNS has three main tasks using three types of nerves. Motor nerves connect to muscles and allow us to move. Sensory nerves are connections to skin and the body that allow us to feel. Dr. Marilène Oliver's image conveys the extensive outreach offered by our sensory nerves to our surroundings (cover image, Fig. 1). Autonomic nerves "automatically" keep us functioning without our having to think. The beauty of nerves is how they act as highways of communication. They share similar but very unique and well designed biology that allows them to transmit their messages. Yet when "nerves" fail, it can be catastrophic. Damage to nerves is called "neuropathy" or "polyneuropathy" if many nerves are involved. However neuropathies may be the most common problem you have never heard of. For good reason, most of us routinely learn about other kinds of neurological problems like stroke, dementia. How about Guillain-Barré polyneuropathy or amyloid neuropathy? Yet Guillain-Barré polyneuropathy is a dramatic condition involving complete paralysis over just a few days in an otherwise healthy person. Recovery can occur but it might take months. Imagine what this might do to your life. Amyloid neuropathy is a relentlessly progressive problem that causes pain, loss of sensation and enormous disability. The first person I ever met with this problem had lost both large toes and several fingers to the condition. These conditions are most definitely "below the radar".

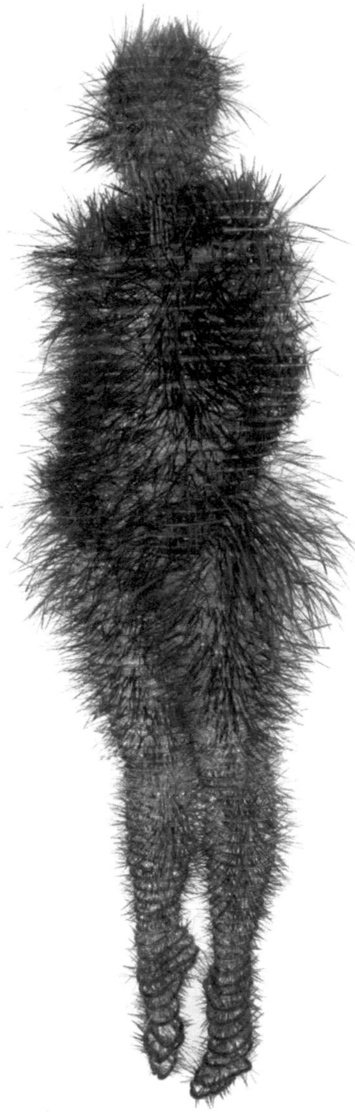

**Fig. 1** "Shredded" by Marilène Oliver, Faculty of Arts, University of Alberta (Reproduced by permission). Oliver reported that the inspiration of the piece arose from a sense of self that followed her learning of the tragic death of a friend. The image conveys the connection between the physical body and its external environment, conveyed by sensory nerve fibers that transmit it (original is laser cut acrylic, MR scans printed on film, fishing wire and crimps. 175 × 50 × 50cm).

## The body brain connection: A fully infiltrated body

Nerves are the essential connections between the CNS (the brain and spinal cord) and the body. They represent the "brain body" connection and can be thought of as an "ome". Like the genome that includes all of our genetic output, the proteonome that includes all of our proteins, the metabolome that constitutes all of our metabolic products, the PNS is a "connectome". Connectomes have also been coined to describe connections within the brain, not our meaning here. In our context, the PNS connectome links the CNS with skin, bones and organs like the gastrointestinal tract, liver, heart and lungs. All of our sensations are coded and transmitted to and from the CNS by peripheral nerves, here called "nerves." The only exceptions are the optic nerves for vision and they are instead considered part of the CNS; in reality they are extensions of the brain to the eyes. However peripheral nerves are different. Wiring diagrams that illustrate the major nerves, as in Fig. 2, vastly underestimate their wide distribution. Nerve endings are microscopic connections that permeate virtually every tissue of the body. For perspective have a look at a magnified image of the external ear of a mouse, with axons labeled by a fluorescent marker (Fig. 3A). The branches are extensive! In the skin, they are only separated by a few microns. A micron is one thousandth of a millimeter in length! For yet greater perspective, examine the fine nerve endings of the epidermis of the skin taken from a human volunteer (Fig. 3B). Finally, the work of Belle, Chédotal and colleagues, that labels individual nerves and branches of the human fetal hand offers a spectacular appreciation of nerve development using state of the art imaging techniques (Fig. 4).[1]

Hence, a map of the human body, based solely on where all of our axons and nerves reside, would outline our external contours and organs very well! It would also be surprising. There are some empty areas. For example, large areas in the brain and liver are surrounded by nerves but not penetrated by nerves. Thus we cannot "feel" ourselves thinking about something. We cannot "feel" strokes or bleeding developing in the brain, unless the covering of the brain, known as the meninges, are stretched. We might recognize that a stroke is occurring because of a change in speech or paralysis but the only sensation triggered may simply be a headache. Rises in intracranial pressure, or pressure within the skull cause headache not because of their damage in the brain but because of meningeal stretch. Our nerves do spread or "ramify" through the meninges and surround its blood vessels. These nerves are "peripheral" despite their location within the skull

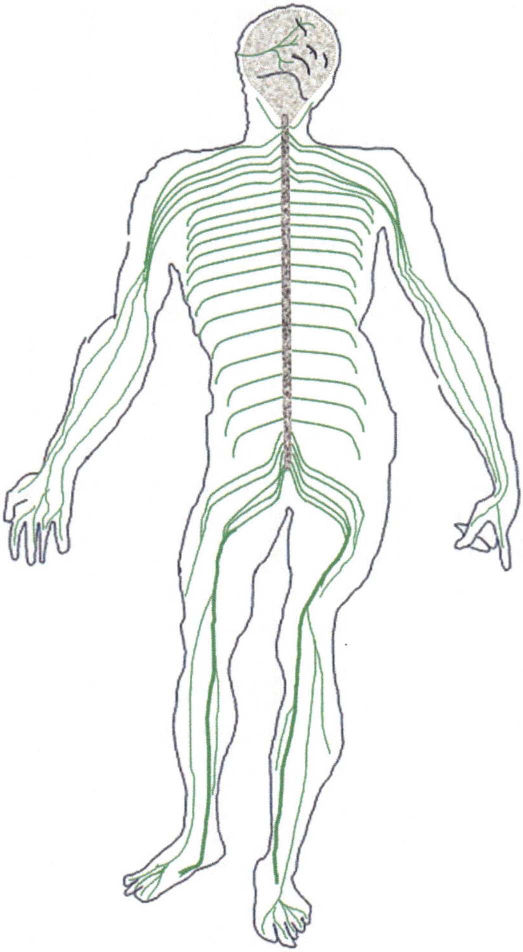

**Fig. 2** A simplified depiction of major nerve trunks in the human body (green). This image does not capture the complexity of fine detailed innervation of the skin or organs innervated by individual axons.

and in the meninges but they do not enter the brain. Instead, the brain itself is composed of CNS neurons and their connections oversee our thought processes. Similarly, nerves do not infiltrate the liver but are found in the capsule covering the liver. Thus, when a person has a needle biopsy of their liver, pain arises as the needle passes through and irritates these surrounding capsule nerves. How did organisms decide where or where not to send nerve terminals? This is a good question, but we can only guess at why we want nerves in some places but not others.

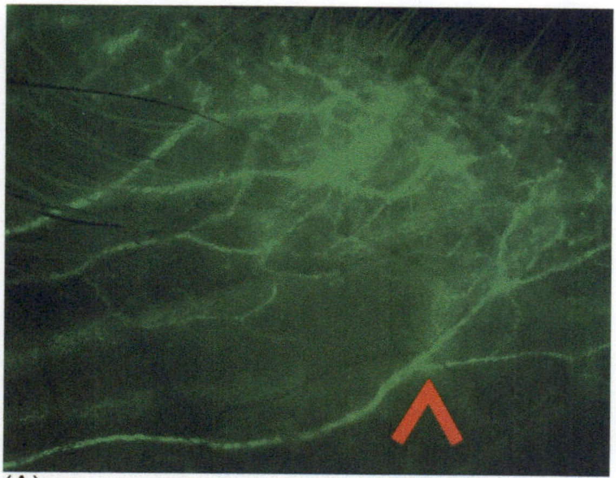

(A)

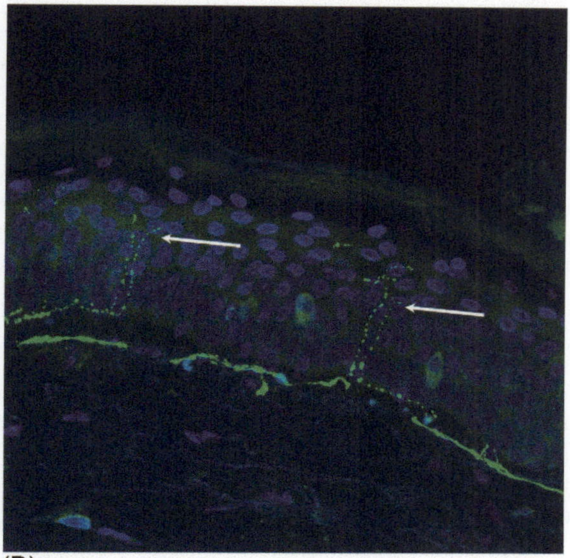

(B)

**Fig. 3** (A) Nerve fibers in the pinna (external portion) of the ear of a mouse engineered to express a fluorescent protein (thy1-YFP) in its peripheral nerves. The fine meshwork of branching sensory axons (arrowhead) extends out to the edge of the ear visible through the skin and hairs. The edge of the pinna is to the upper right (Image taken by Virginia Woo, Zochodne laboratory). (B) A section of human skin from the lower limb of a normal subject from a small punch biopsy. The image depicts the layers of the skin with keratinocytes (skin cells) having blue nuclei in the epidermis and fine green fluorescent epidermal axons (arrows) labeled using immunohistochemistry to a nerve protein called PGP9.5. The lower left is the deeper dermis showing a bundle of nerve fibers that supply the branches to the epidermis. *(Image take by Dr. Jesi Bautista, Zochodne laboratory.)*

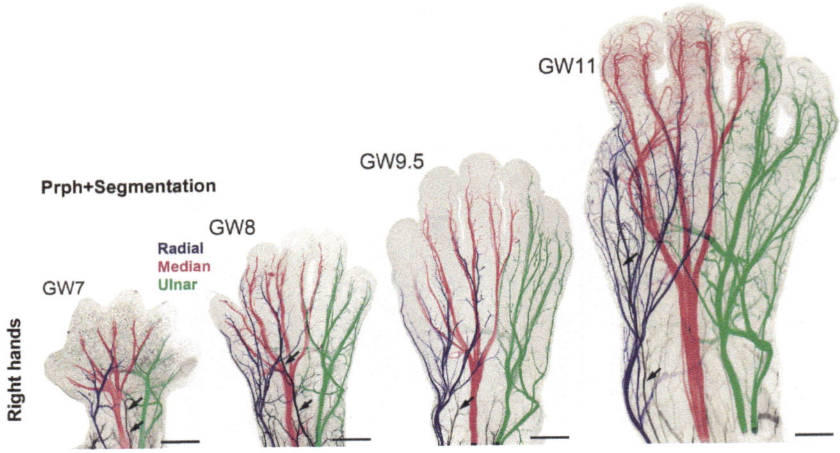


**Fig. 4** Images of developing nerves in the human fetal hand from gestation weeks (GW) 7-11. The images were created from solvent cleared (renders the hand transparent) tissues. The images show radial (blue), median (pink) and ulnar (green) labeled nerves, as they grow into future digits (fingers). The arrows point to a small branch (black) of the musculocutaneous nerve, transiently entering the hand (not found in the adult hand). To label the nerves, the authors used immunohistochemistry directed to a nerve protein called peripherin and performed reconstruction of the pathways of the nerve in 3D. For further details including spectacular images and videos please refer to Belle et al (Cell 169: 161-173, 2017).[1] *(The images were reproduced with the kind permission of the senior, corresponding author Dr. Alain Chédotal.)*

Ignoring the spaces that lack nerves, with their degree of infiltration, nerves are among the most common structures in the body. However, despite their abundance, we know less than we might. How do nerves decide exactly where to locate? Is it by chance or is there a system of rewards and punishments that will determine where they grow? "Rewards" are sometimes called "attractant cues" by biologists because they encourage nerve fibers to come hither. "Punishments" in contast, are called "repulsive cues" because they tell the fibers to go away. Think of attractive and repulsive foods or situations and you will understand. Future research about nerves may eventually answer these fascinating questions. Another question is whether nerves are "fixed" and immobile like concrete highways or constantly growing and remodeling like hiking pathways? You may be surprised to know that they are actually highly plastic and mobile! Perhaps then nerves are more like hiking pathways! This is likely true for the fine endings of nerves. However large nerve trunks like the sciatic nerve in the leg or median nerve in the arm are more fixed in the body. These large nerves are more like interstate or provincial main thoroughfares. Have you driven on highway "401" in Ontario, a large freeway that connects Toronto, Montreal and Windsor? Think of the sciatic nerve being its counterpart in the leg connecting the spinal cord to the leg and foot. Of course, my argument is that the nerve is a far more interesting place to travel than this famous but somewhat tedious Ontario freeway.

The nerve-immune system explores how peripheral nerves and immune cells work together. A number of ideas, only some proven, around this relationship abound. For example, nerve endings, or terminals can release molecules from their endings. These in turn can activate or signal immune cells. The immune cells in turn, can release further molecules that either worsen or dampen inflammation. Given a pathway from the brain through nerves to the immune system, its possible that the immune system could be "instructed". The brain might signal immune cells and direct their responses. It is an interesting connection that has had plenty of attention.

Some of the ideas around the brain immune connection are unproven and others remain controversial. For example its thought by some that special signals through nerves from the brain might alter your level of immunity. An example is depression, and the idea is that a change in brain function might lower your immune function. While there is some evidence for this, there are also instances where autoimmune disorders involving an overactive immune system are instead activated by depression. Another example might be the idea that positive thinking, transmitted through your nerves might prevent or suppress cancer. This idea is without much support. Unfortunately this popular idea nonetheless can make persons with cancer feel guilty that they have failed to think positively enough. Like all of science there are good ideas out there. Some ideas center on unexpected relationships within our complex connectome. However some ideas should be shelved until there is solid science behind them. What we do know is that a fully integrated system of nerves neighbor virtually all other cells of the body. This does imply that nerve and body health are closely related.

What does it take to activate nerves? Under "normal" conditions, without thinking about your body, nerves are active. How do we know this? A procedure known as "neurography" allows investigators to record ongoing nerve discharges with fine wires in nerves. What neurography tells us is that "resting" discharges are common in nerves, but there are fewer discharges in a normal "resting" nerve than if they are signaling something terrible, like pain from a serious injury. What we also know is that the brain has an important role in filtering this ongoing flood of information from nerves throughout the body. Only the most intense, localized and persistent information reaches awareness. This is obviously very helpful for allowing us to focus on our tasks at hand without annoying and ongoing distractions. Most of the time we are unaware of isolated and transient nerve discharges. They are minor blips in a universe of noise. However, when such discharges arise from a number of adjacent nerves together, suddenly they might be important. The signal rises above the level of ongoing "noise". For example, many of us rarely pay attention to how many parts of our body send us discharges at any given time. Our bodies change constantly and these changes are continuously detected and signaled by nerves. These might include simple actions like sitting, standing or walking. Being aware of every nerve discharge that erupts because of our posture or stance would overrun our brain with noise! Filtering is important! This is not simply a matter of attention. Attention requires active engagement of the cortex of the brain and determines what we choose to focus on.

The purpose of this book is to raise much needed awareness. However it is also about much more than awareness. There is a beauty in the structure and function in nerves that is worth fully appreciating. After some terminology, some anatomy and some physiology, its splendor unfolds. Not only is the operation of the PNS elegant under normal circumstances, but its behavior during disease is fascinating and how it regenerates is exquisite. It may be that the reader is planning a career in health sciences or has an acquaintance pursuing this goal. Others may have had to deal with neuropathies themselves or in family members. Yet others I hope to convince are simply interested in unique biology, called here "peripheral neurobiology".

# CHAPTER 1

# Elegant wiring: Structural beauty of the peripheral nervous system

## How are nerves structured

Nerve fibers mingle among the cells of almost every tissue in the body. They are more widespread than blood vessels. When a tissue has its normal and expected group of axons, it is "invested" with axons or "innervated." Tissues that lose their nerve supply are "denervated." There are some surprising exceptions as described in brain and liver. Of course, the scalp and skull are also innervated, explaining why it is uncomfortable to bang or injury your head! Damage to the liver does not usually cause pain unless the innervated liver capsule is stretched or irritated. If you drink alcohol heavily, you may not be aware that your liver is undergoing damage because you do not feel it. Perhaps if you could, alcoholism would no longer exist as a public health problem!

The nerve trunks, also just called the "nerves," are specialized superhighways for axons that transmit signals. Nerve trunks contain hundreds to thousands of individual axons, our "wiring," along with supporting cells, blood vessels, and proteins that add structure and strength. Each nerve contains fascicles, the contents of which are called the endoneurium, compartments that house axons, ("inside" the nerve).[2] Certain large nerves may have 20 fascicles, others 1–2 and these may identify groups of axons with specific destinations in some nerves, like the large sciatic nerve in the upper thigh. The axons in each fascicle are also diverse, with admixed motor, sensory and autonomic types that sometimes change fascicles as they move toward their target destinations. Some further definitions will help (Fig. 5). The perineurium ("around" the nerve) is a layer of interwoven cells that surround the all important endoneurial fascicles. They offer protection of axons from the elements by erecting a "blood nerve barrier," preventing free access of blood or other body fluids to the endoneurium. Many nerve disorders involve breakdown of this important barrier, allowing vulnerability of axons to inflammation and other difficulties. The epineurium ("on top" of the nerve) links and surrounds all of the fascicles binding them into

*Our Wired Nerves*
https://doi.org/10.1016/B978-0-12-821487-9.00001-5

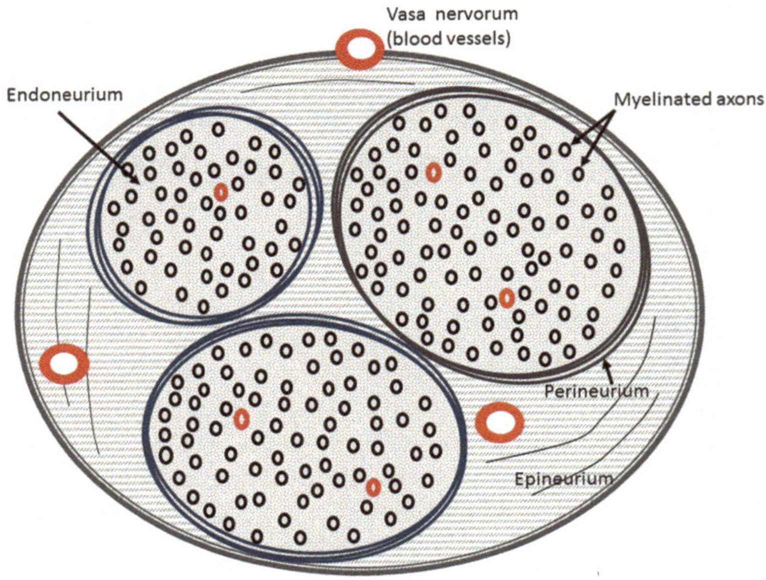


**Fig. 5** Diagram of a cross (transverse) section of a peripheral nerve trunk with three fascicles illustrating the outer epineurium, perineurium surrounding the fascicles and endoneurium containing axons within them. Smaller unmyelinated axons are not shown. Vasa nervorum are the blood vessels that supply the nerve trunk.

a cohesive structure or "nerve trunk" with connective tissue. It includes collagen, an important supporting protein, arterioles and venules that ultimately supply blood to the endoneurial compartment. These blood vessels are called "vasa nervorum," the blood supply to nerves that mingles through the epineurium in complicated arrangements (a vascular "plexus"), sending feeding vessels into the fascicles. Most of the blood vessels within the endoneurial fascicles are capillaries, also tightly constructed to add to the blood nerve barrier. In some nerve disorders, this vascular supply is targeted by disease. An interesting twist on this story is that nerve trunks, the highways of axons that travel from the spinal cord throughout the body, are self innervated. In other words, they are "invested" with small axons of their own. This will be described in more detail later. The overall structure of nerve trunks is maintained through much of their travels in the body until they reach target tissues, or connect to the nerve roots and spinal cord.

The "wiring" or axons of the PNS have structure similar to those found in the brain but they have vastly different distributions and actions. This can seem confusing but some anatomy will help. Axons in general are thought of as "wires" but are actually microscopic tubular structures that arise from the nerve cell body, also called a perikaryon. The perikaryon is a roundish or oblong structure with a nucleus and when considered together with its axon is of course called a neuron! All parts of the neuron are surrounded by a cell membrane that accounts for its shape. While surrounding the axon, the membrane also facilitates information transmission. However unlike a true wire, where electricity travels through a full copper or silver metal structure, the axon uses a layer of itself, called axoplasm, just beneath its tubular membrane to transmit electric signal. The portions of axoplasm further within have other roles. Not surprisingly, axons are very specialized structures.

The contents of axons differ from any other cell type. Longitudinal ropes of proteins known as neurofilaments and microtubules are found within the axoplasm (Fig. 6). These structures allow the axon to connect over long distances without falling apart, like a long telephone cable. In fact, axons extend from millimeters to meters away from their origin in the neuron cell body and may have several branches. Due to their length, axons can be found at some distance from where they start. Axons may be larger or smaller in diameter, and their size is related to their function. The axon is therefore like the essential "wiring" of the cable, but it cannot operate alone. Like a telephone line, it needs insulation and support. Thus, it requires a highly specialized cell partner to work properly, the Schwann cell, described next.

Schwann cells (SCs), named after Theodor Schwann (1810–1882), who was an anatomist and physiologist from Germany, are the intimate partners of neurons and their axons. Without them, axons in the PNS cannot survive. SCs belong to a larger family of cells known as "glial" cells, supporting cells in both the CNS and PNS. Besides the Schwann cells, glial

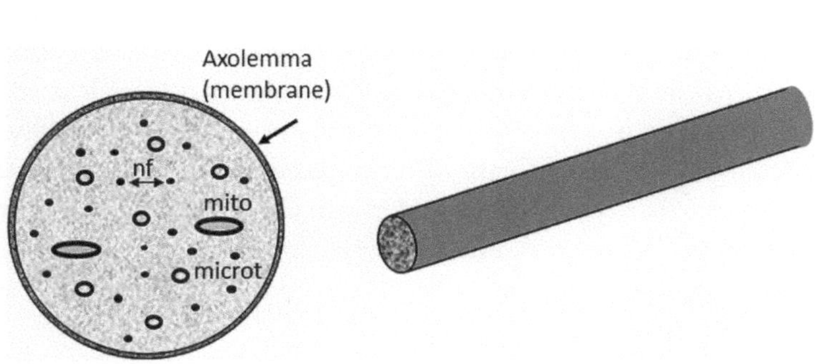


**Fig. 6** Simplified diagram of an unmyelinated ("C") axon transverse section and longitudinal. The axon is surrounded by a single membrane (axolemma). The inside of the axon includes neurofilaments (nf), microtubules (microt) and mitochondria (mito).

cells include astrocytes, oligodendrocytes and microglial cells, none very important for this book! These glial cells exclusively support the CNS! In contrast, SCs, PNS glial cells that make myelin, are not found in the CNS. CNS myelin production, which we will consider below, is taken over by oligodendrocytes rather than SCs and CNS axons are ensheathed by myelin made by oligodendrocytes. Astrocytes of the CNS support and communicate with neurons. It is interesting that the anatomical border dividing SCs in the PNS from oligodendrocytes and astrocytes in the CNS is strict. Using special markers, anatomists have noted that SCs abruptly stop at the junction where peripheral nerves enter the spinal cord. Similarly the CNS glial cells do not migrate into peripheral nerves. What controls this firm barrier is not fully known. It is clear that all types of glial cells have "instructions" and rarely disobey. SCs are the supporting PNS glial cells we are interested in here.

SCs come in several flavors. Their most important work is to ensheath axons with myelin, the insulation surrounding larger axons (Fig. 7). However SCs also communicate directly with axons. Some of this communication involves crucial molecular signals known as "trophic" or growth factors that keep axons healthy and functioning normally. In the reverse direction, axons send signals to SCs. Finally SCs also scavenge or collect and dispose of damaged axons and their debris. They truly are essential partners, including dealing with garbage.

SCs may also actually exchange their parts, or organelles ("small organs"), with axons.[3] While this has only been suggested recently, the idea is that special key complexes, organelles, within a cell are shifted to axons from SCs, mainly a one way exchange. Ribosomes are types of organelles made up of multiple clustered and specialized proteins which when acting together, make new proteins. Another very important large organelle, but one not yet shown to transfer, is called the mitochondria and its function is to generate energy. Where this kind of transfer becomes especially interesting is among inherited disorders of the peripheral nerves, that we will discuss in greater detail later. In some forms of inherited nerve disease, the abnormal mutated protein that causes the disorder actually originally belongs to the SC, not the axon. Despite this, axons nonetheless develop damage as the disorder progresses, suggesting that defective cargo may have been transferred from the SC to the axon.

Unlike adult peripheral neurons and axons, SCs are mobile and proliferate. These properties provide the flexibility that neurons and axons do not have. Some SCs seem to be constantly in flux growing, migrating, dividing

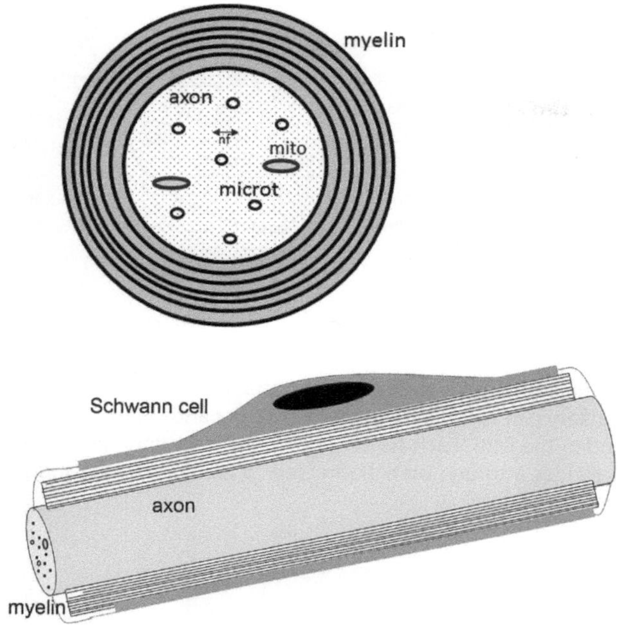


**Fig. 7** Simplified diagram of myelinated axon transverse section and longitudinal. Normal myelinated axons have many more myelin lamellae (layers) than depicted. The inside of the axon includes neurofilaments (nf), microtubules (microt) and mitochondria (mito). Each myelinated axon is associated with one Schwann cell per internode (between each Node of Ranvier).

then dying. Others are more stable, and these are the SCs that synthesize myelin. Among the various flavors of SCs we have discussed are types that surround neuron cell bodies. These are also called satellite cells, and another name for them is "perineuronal satellite cells." "Ganglia" are collections of neurons and glial cells. In the PNS, the cell bodies or perikarya for sensory neurons with their surrounding satellite cells are grouped together in clusters of cells called dorsal root ganglia, or DRG that are found up and down the back, alongside the spine (Fig. 8). The trigeminal ganglion houses the sensory neurons that have branches supplying the face, head and neck. Perineuronal satellite cells are closely applied around sensory neuron cell bodies and while they do not myelinate it, they cover almost all of the neuron (Fig. 9). Their role likely involves critical communication and exchange with the neuron. Yet another type of SC is only found where axons and muscles connect, known as neuromuscular junctions (NMJs). These SCS, known as "perisynaptic" or terminal SCs, are essential for constructing the

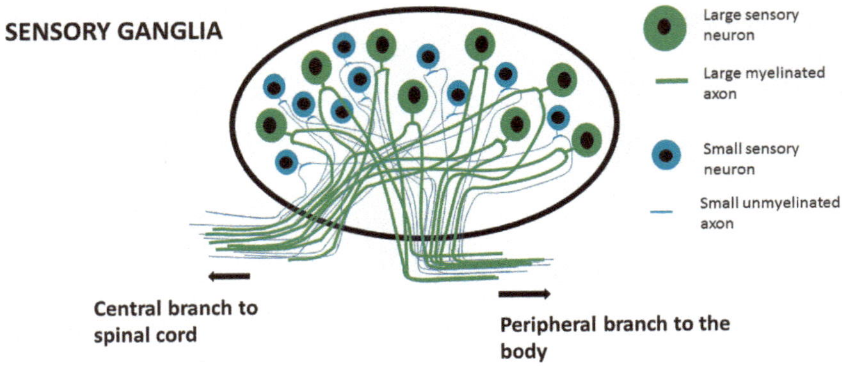


**Fig. 8** Diagram illustrating the structure of a dorsal root ganglia (DRG) that houses the cell bodies (perikarya) of sensory neurons. Both larger and smaller sized cell bodies and axons are found in the DRG. Each sensory neuron is "pseudounipolar" with a short single axon that divides into one branch directed to the spinal cord and another to the periphery.

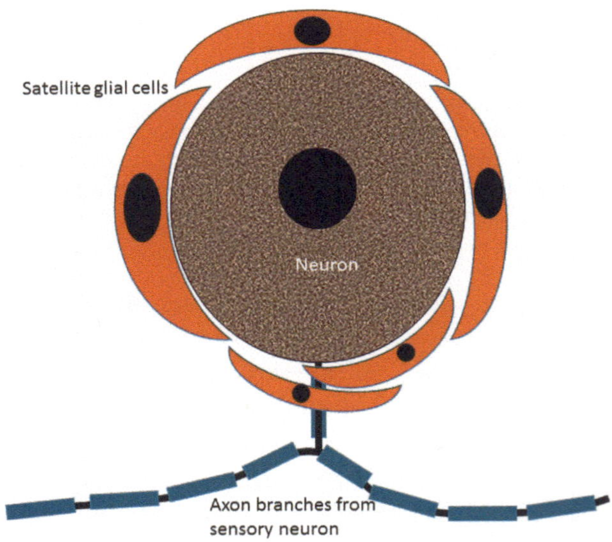


**Fig. 9** Illustration of a DRG sensory neuron surrounded by satellite glial cells and one output axon that divides into two branches.

architecture and ensuring the function of the junction. They are found at the "terminal" end of the motor axon, but they are not about to die!

Small axons lacking a myelin sheath are by far the most numerous axons in the PNS and they transmit sensory information or autonomic signals. They are also sometimes called "C" axons/fibers by physiologists, a historical title that was given when recording their signals in a dish! More simply they are called "unmyelinated" axons. Unmyelinated or C fibers are clustered together in bundles with SCs together termed "Remak" bundles named after a German/Polish neurologist Robert Remak (1815–1865) who described them. Robert Remak and his mathematician grandson had lives that were distinctive for other reasons as well-both suffering persecution, an issue that clouded many spheres of science in the last century (https://en.wikipedia.org/wiki/Robert_Remak).

A Remak bundle is a collection of several C fibers held together by one distinctive SC (Fig. 10). These SCs specialize in partnering and collaborating with small axons in a very different way than myelin forming SCs. The individual fibers are fully surrounded and separated by extensions of the Remak SC cytoplasm and when viewed, the bundles are elegant, seen best on electron microscope (EM) images. Each bundle can contain as few as 1–2 axons or beyond 20, all individually wrapped. This form of simple "wrapping" however is not myelin, but arises from single SC processes. Remak bundles represent a fascinating snapshot of how two different and interdependent cells, the axon and SC, share an intimate relationship. Another interesting fact is that the individual axons within a Remak bundle, despite looking the same to the anatomist, might in fact be headed to completely different directions, or targets. Any given unmyelinated axon may be part of one Remak grouping, but as it diverts to a different target, its neighbors change. Overall, despite their different functions and targets, autonomic and sensory C fibers look identical under EM. In defense of elegant EM studies

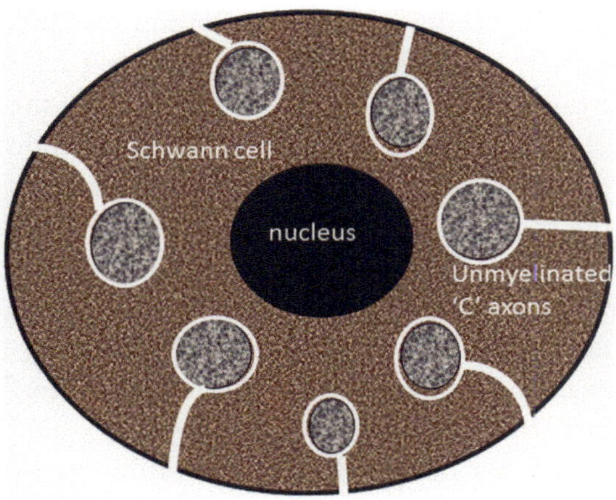


**Fig. 10** Simplified diagram of a Remak bundle, a Schwann cell associated with several unmyelinated axons.

are their ability to provide anatomists with what is called the "ultrastructure" of a tissue such as small axons and SC partners. Ultrastructural views of axons, in turn, open another dimension of their biology and this will be discussed later. An important point is that it is the information that C fibers carry, not their fine ultrastructural anatomy, that tells us their role.

Among the axons within Remak bundles, unmyelinated "C" axons are the chief highways of the autonomic nervous system, or ANS. The ANS is a critical part the nervous system, discussed in more detail later, that maintains automatic (autonomic) body function. Temperature regulation is controlled by nerves causing blood vessel dilatation (expansion) that releases heat or constriction (narrowing) that prevents heat loss. Activation of autonomic nerves causes sweating. The ability to move one's bowel or empty the bladder requires these nerves. Autonomic axons control gut motility and are essential for sexual function. Even more subtle functions, such as the control of heart rate variation, also depends on autonomic nerves. If you use wearable technologies that provide heart rate recordings, you will know that hear rates can vary quite extensively, even with quiet breathing. Autonomic nerves signal to the heart whether to speed up or slow down the heart rate. Also in Remak bundles, small diameter C sensory axons transmit information from the skin and organs. These may include pain, cold or heat sensations. More on this later.

Larger axons are less numerous than C fibers and have a surrounding sheath of myelin made by their SCs (Fig. 7). Myelin is a form of very organized wrapping that winds around axons in layers like an insulator in a beautiful example of structural regularity. It is mainly composed of lipid (fat) that comes from SC membranes but within this lipid key proteins are placed strategically that gives myelin its character. Each wound layer of myelin, together called lamellae, is simply an extension of the SC cell membrane. Most myelinated axons have between 50 and 150 individual wraps of myelin around them.[4] How is this accomplished? Each wrap or layer is laid down because the motile SC is able to move around the axon to myelinate it. Multiple wraps are then compressed to form a compact sheath. Dr. Patricia Armati, a pioneer Australian neurobiologist in the study of myelin amused her audiences by asking whether myelin winding happened in a reverse direction in Australia, in the Southern hemisphere![5] I still do not know the answer to this, but expect not.

Myelin is essential for nerve impulse transmission or propagation, discussed below. Myelin itself is similar in the CNS and PNS but with some important exceptions, including the cell that makes it-SCs in the PNS and oligodendrocytes in the CNS. While the lipid part of myelin is similar, the proteins associated with myelin differ in the PNS and CNS. This distinction is important because it explains why some diseases of myelin are very different in the CNS compared to the PNS. For example MS involves inflammation of CNS myelin only and is strictly a disorder of the brain and spinal cord. In contrast PNS myelin inflammation is associated with disorders such as Guillain-Barre syndrome. In either case, loss of the myelin sheath around groups of axons is known as "demyelination." Demyelination in the brain causes neurological symptoms problems that are unlike those associated with demyelination of nerves and their treatments are very different.

Along the length of all myelinated axons, myelin sheaths form segments approximately 0.5–2.0 mm in length.[4] These resemble a rosary or string of beads, except that the "stones" or "beads" are elongated myelin segments and the parts that separate them called nodes are very short (Fig. 11). The segments of myelin that surround axons start and end at junctures known as "Nodes of Ranvier" named after another pioneer, a physician and anatomist, Louis-Antoine Ranvier (1835–1922) from France. "Demyelination" occurs in very distinctive patterns including loss involving individual segments between the nodes, "segmental demyelination." Another form of demyelination is called "paranodal demyelination" in which myelin is retracted from around the nodes, baring parts of the axon normally insulated by the myelin sheath. In both forms of demyelination, loss of myelin insulation changes the electrical properties of the axons. Their ability to take up ions is altered because proteins known as ion channels are displaced from where they belong. Ion channels are protein "pores" in the membrane of the axons that allow charges to cross it. They maintain the electrical function of axons and allow it to transmit signals. Since the role of the node is to help propagate an electrical signal, loss of myelin either in a specific segment or around the nodes alters normal signal transmission.

Electrical signals or "impulses" that travel or propagate along axons are called action potentials (APs). However, how they do this differs between

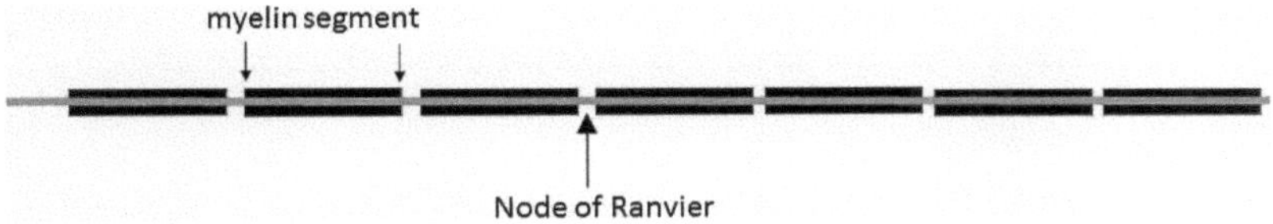


**Fig. 11** Diagram illustrating a myelinated axon with segments of myelin separated by Nodes of Ranvier.

axons with myelin and the C fibers that lack myelin. In an axon without myelin, action potentials travel slowly and deliberately. All of this differs in myelinated axons that have developed a unique short cut for moving their signals; it is considerably faster and directs current from node to node. We will discuss this in more depth below. Nevertheless, this physiological process allows much more rapid conduction than its less impressive C fiber neighbors. Unmyelinated "C" fibers have slower "continuous" conduction without nodes to enhance the conduction circuits. Rapid conduction is important for function: muscle movement, light touch sensation, feeling vibration, recognition of the position of limbs and sensing some types of pain.

Myelinated axons can also be subdivided into several types (Fig. 12). Large myelinated motor axons are termed alpha motor neuron fibers and they control voluntary muscle movement. Smaller diameter motor fibers known as gamma motor fibers help regulate how sensitive muscle fibers are to being stretched. Larger sensory axons are myelinated. Finally there is also an intermediate class of medium sized sensory myelinated axons known as A delta that transmit pain signals.

The origin and terminations of nerves are as important as what the "wires" look like. Despite trying to convince you that PNS and CNS neurons are very different, there is actually overlap. For example, motor axons that control muscle movement, originate from neuron cell bodies in the CNS, the spinal cord. In particular, they arise in a portion of the cord known as the ventral or anterior horn that is part of its gray matter. Recall that in the CNS, gray matter refers to structures that contain cell bodies of neurons whereas white matter consists of bundles of connections, or axons.

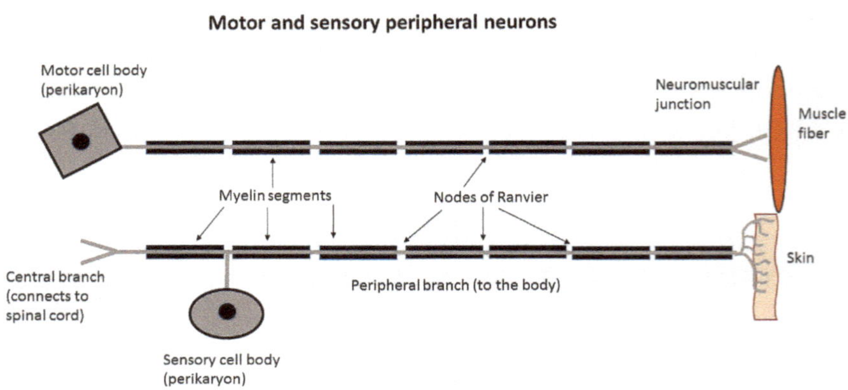


**Fig. 12** Diagram illustrating myelinated motor axons and sensory axons with myelin segments and Nodes of Ranvier.

Motor neurons, originating in spinal cord gray matter, thus have parts both in the PNS and in the CNS. The large motor neuron cell bodies or perikarya are also interesting in that they are hexagonal in shape and receive extensive inputs through processes called dendrites. For many CNS neurons, wide branching dendrites allow them to receive input from many other neurons through synapses, the critical sites of connection between neurons. In summary, and unlike PNS sensory neurons in ganglia that we will hear about, motor neurons resemble other CNS neurons with large dendritic trees, allowing them to be controlled from multiple directions. These extensive dendritic trees moreover each contain yet smaller processes known as dendritic spines. All of these distinctions indicate different biology. For example, motor and sensory neurons have different susceptibilities to diseases. In motor nerves, axons of other neurons, for example in the cortex of the brain, descend to make contact with the motor neuron dendrites. Since many types of signals are sent to them, dendrite antennae are required, far beyond where the cell body sits. Indeed, beautiful anatomical studies by Ken Rose and colleagues at Queen's University in Kingston, Canada showed that the input into a single motor neuron can occupy much of the anterior horn of the spinal cord.[6] These connections, in turn, are essential to transmit motor commands from the brain. To see the full tree of motor neuron connections within the gray matter of the spinal cord is captivating. The possibilities for connectedness and input are striking to contemplate (Fig. 13).

Unlike motor neurons, sensory neurons have their own place of residence outside of the spinal cord (Fig. 13). Why the body should be built with completely different homes and structures for motor or sensory transmission is something to wonder about. Sensory neurons have their cell bodies housed in structures called ganglia, mentioned earlier. For nerves below the neck, these are known as dorsal root or spinal ganglia, or DRGs for short. For cranial nerves that supply sensation to the most of the face, mouth, tongue and head the cell bodies are also in similar kinds of sensory ganglia, in special locations. An example is the sensory ganglia of the trigeminal nerve is also known as the 5th cranial nerve. This nerve is complex in that it supplies motor nerves to muscles for chewing, sensation to the face and inside of the mouth and sensation to the tongue. It provides what is called general sensation to touch, pain and cold. Yet other nerves and ganglia are responsible for signaling taste from the tongue and mouth. Taste sensation is called a "special sensation" and its nerves travel with the 7th and 9th cranial nerves. The 7th, or facial nerve supplies taste to the front two thirds of the tongue, whereas the 9th or glossopharyngeal nerve, supplies the posterior two thirds. More about taste later!

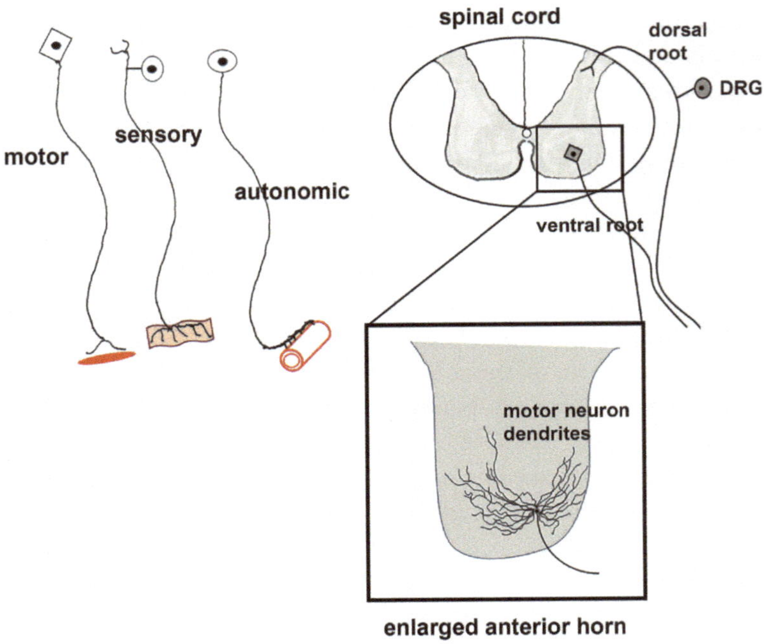


**Fig. 13** Diagram illustrating motor, sensory and autonomic axons. The connections of sensory axons and motor axons with the spinal cord are shown. The enlarged inset diagram illustrates the extensive dendrite receptive "tree" of a motor neuron, based on work published by Rose and Cushing.[6]

The trigeminal nerve has its own story and is worth a digression. It is well known to clinicians for the problem known as "trigeminal neuralgia." This is an intensely painful affliction said by some to render pain more severe than natural childbirth. The word "neuralgia" in general refers to conditions that inappropriately activate pain signals from sensory axons. Trigeminal neuralgia often arises from pressure on the nerve from an extra loop of a nearby blood vessel and separating them surgically relieves the condition. We will discuss this in more detail later.

Motor and sensory axons arising from their cell bodies have different pathways into the nerve trunks of the body (Fig. 13). Motor axons exit from the front, ventral or anterior horn of the spinal cord to form ventral nerve roots. DRGs housing sensory cell bodies are found adjacent to but separated from the spinal cord. "Central" branches of sensory neurons from DRGs connect to the spinal cord through its back or dorsal side through dorsal roots. "Peripheral" branches of the DRG send their sensory axon branches out to the body. This overall arrangement allows DRGs to send one branch to the spinal cord, so that information can travel to the brain, and other to the body, where they can receive information. Ventral motor and dorsal sensory roots eventually join up to form what are called mixed spinal nerves. Mixed spinal nerves then connect through a spaghetti, or freeway interchange of intermixing nerve bundles, called plexi, that eventually form major nerves. Because of this arrangement, virtually all of the nerves destined to the body contain both motor and sensory axons.

Using modern techniques to label and identify motor axons, Drs. Gesslbauer and colleagues from Vienna did something no one else had attempted.[7] They took samples from human arm nerves from their origin in the spinal cord to their branches out into the hand. As the nerves leave the spinal cord in the neck to travel to the arm, the complicated set of bundles entering the arm is known as the brachial plexus. Recall the term "plexi" above, referring to a complex of nerve interchanges as the mixed spinal nerves form peripheral nerve trunks. In essence the brachial plexus rewires axons from the spinal cord so that they can go to specific muscles. For example, axons from the 7th cervical nerve root eventually travel to the triceps muscle that normally extends the elbow. In doing so, they take several nerve highway interchanges within the brachial plexus. If this were you and I on a travel expedition, we would definitely need our trip navigator to get us to the right destination. From the spinal cord at the 7th ventral nerve root level, motor axons join sensory axons in the mixed spinal nerve, then pass along through what is known as the middle trunk of the brachial plexus, through the posterior division, then through the posterior cord of the brachial plexus and then into the radial nerve finally and eventually reaching the triceps muscle. What a complicated route!

How did Gesslbauer and colleagues collect these samples? They had special permission from organ donors allowing analysis before the nerves degenerated after death. The approach is known as immunohistochemistry and allows the use of a special stain to highlight exactly what types of axons were in any given sample. Recall that axons may look alike, but have very different roles. Immunohistochemistry (IHC) is a way to label the tissue for specific proteins, using antibodies. Recall that antibodies are proteins made by the immune system to attach to very specific targets. Targets for an IHC stain might be proteins only found in motor axons or in sensory or autonomic axons. The anatomists stained for ChAT, or choline acetyl transferase, an enzyme only found in motor axons to help sort out how many of the axons were motor, or went to muscles. The answer was surprising to many. In fact motor axons, sometimes clustered into some specific parts of the nerve, accounted for only around 10% of the axons they could see in these specimens. All of the rest were sensory and a small number were autonomic axons. This was unexpected in suggesting that of our wiring, about 10% of it is designed for outgoing messages. Up to 90% of our wiring is designed to receive incoming messages, of all types. This is itself a remarkable filtering: plenty comes in, but selective signals come back out. A final interesting outcome of this study was that the parts of the body requiring very fine motor control, such as the hand had no higher number of motor axons than big muscles that make large movements. The fine control of hand muscles must depend on precise timing of how their 10% motor axons can fire, modulation that arises from our brain. This is my concession to the idea that the CNS is after all important, if only to make sure that PNS fibers operate correctly!

We have mentioned DRGS, the ganglia where the cell bodies of sensory nerves reside. DRGs are striking and attractive compilations of cells of varying sizes. In humans, they are challenging to dissect out through the spinal bones that surround them. As a result, human DRG studies are comparatively rare and they are less frequently sampled in autopsies. For obvious reasons, DRGs are also not often biopsied in living persons. However we do know important details about DRGs. Within their three dimensional capsules the neuron cell bodies can often be found layered in the outer portions but feeding their branches toward the center. Additional neurons are spread through the center of the DRG. Sensory neurons that are larger have different roles than smaller ones and each type has a different assortment of molecules within them. Large sensory neurons in the DRG support larger axons in the nerve. Larger axons transmit information about light touch, vibration and position. Smaller sensory axons transmit information about pain and temperature. The axons that exit from cell bodies collect within the center of the DRG then divide into their two branches, peripheral to the body and central to the spinal cord (Fig. 8).

The complexity and beauty of DRGs becomes evident using IHC stains that highlight their specific proteins. IHC uses an antibody that attaches itself to a protein or peptide, which is then labeled with a fluorescent or other dye. This fluorescent label may already be part of the original antibody, or the label may be on a second antibody directed against the first, a more common and economical method. Thus, the first antibody binds to the protein or peptide of interest of the neuron and the second binds as a label attached to the first antibody. The fluorescent tag highlights where in the cell or in which cells the target is located. One can use several different antibodies and fluorescent tags and generate remarkable images of how proteins are distributed among the DRG neurons. Combined with this may be other fluorescent stains that label all of the nuclei of the cells in the DRG. A related IHC technique, called immunoperoxidase instead uses a nonfluorescent tag that labels the positive structures as brown. These images can also be highly informative but they are not quite as remarkable as the fluorescent landscape!

IHC illustrates how the large and small sensory neurons in DRGs are unique by highlighting their individual roles and destinations. For example, large neurons stain heavily for a protein known as neurofilament, a part of the backbone or "ropey" lattice structure of the nerve. Within the cell body, the intertwining complexity of neurofilament strands is conspicuous and quite beautiful. These proteins are originally synthesized in the cell body but then are exported, or pushed out into the axons. Upon their arrival in axons, they become highly organized and are carefully laid out like railroad tracks (or lines of ropes) giving structure and resilience. Other proteins identified in peripheral neurons, such as growth factor receptors, reveal their biology. The type of growth factor receptor protein that is found on a given DRG neuron indicates how it might respond to injury. For example, small neurons have less neurofilament staining. However among this smaller sized neuron group, there are yet further differences and two main types. One classical type contains the pain peptide SP (Substance P [SP]) that we will consider later. These also harbor the receptor for nerve growth factor (NGF), an important protein discussed later, known as TrkA. A different type of DRG neuron is called IB4 because of a unique molecule, "lectin," on its cell surface. Both of these smaller sensory neuron subpopulations help to transmit pain. Since recognizing pain is essential for survival, multiple pain pathways exist. This includes our two similarly sized, TrkA and lectin, but molecularly different types of sensory neurons! Whether these two types of neurons have identical or unrelated roles in pain sensation is not known.

Some additional explanation is required of the unique wiring of sensory neurons and axons. Recall that sensory neurons do not have dendrites and dendritic spines. Since sensory signals require precision, wide signaling using dendrites might be disruptive and it is difficult to know where they might fit into the sensory neuron tree. Sensory transmission requires specific and sensitive axons attuned to different stimuli with precise connections. They must be able to tell the nervous system where and what sort of sensory signal is heading toward the brain. Thus "hard wired" single sensory neurons require a single minded purpose, and the brain can later parcel out these inputs and interpret them as needed. If we reach to swat away a mosquito that has landed on our arm, it is essential to have a precise "map" of where it has alighted allowing a rapid motor response (swat!). Given all of this, we note that sensory neurons have what is called a "pseudo-unipolar" shape. From each cell body or perikarya, one process emerges (unipolar), that then branches into two, and only two axons; hence the term "pseudounipolar" is used. The emerging processes are output axons without dendrites, the peripheral branch assembled together with motor axons travel with the peripheral nerves to the body. The central branches travels from the DRG through the dorsal root to the spinal cord. This branch then enters the dorsal spinal cord and connects to pathways traveling up to the brain. The point at which the central branch exactly enters the spinal cord has a very imaginative name-the dorsal root entry zone (DREZ)! The destination zone in the spinal cord for dorsal roots is called, of course, the dorsal horn of the spinal cord.

Sensory signals thereby travel from the skin, up the nerve to the ganglia through its peripheral branch. At the level of the ganglia the impulse then travels seamlessly into the spinal cord by next invading the central branch. It does so without required travel through the perikarya in the DRG, like a bypass that avoids a metro city center. It's a thoroughfare or freeway without interruption. The DRG is a way station, not unlike a truck stop on an interstate highway. It's a detour off of the main highway with a single ramp, making it "unipolar."

In the dorsal horn of the spinal cord where the dorsal roots connect, relay centers for sensory information are found. Here the axons from sensory neurons are now within gray matter and form synaptic connections with spinal cord neurons. These neurons may in turn send their own branches to travel up and down the spinal cord. However, let us return to the incoming branches from the DRG sensory neurons. Another name for incoming sensory axons is primary "afferent." Afferent means inward (into the spinal cord). It refers to a signal heading toward the brain and cord from the body. "Efferent" means outgoing. Motor axons headed out are efferents. Sensory signal pathways become yet more complicated when they enter the spinal cord. Sensory incoming afferent fibers thus connect to some spinal cord neurons called interneurons. As their name suggests, these connect from neuron to neuron. Here is an important change. Although sensory signals reach the spinal cord by traveling along a single axon, within an uninterrupted nerve interstate highway, they now are held up by traffic lights. Unlike a straight wiring plan, there are now multiple relay stations when the signals reach the dorsal horn. Interneurons, the pieces of this relay system, can enhance or delay the traffic flow by applying local excitatory or inhibitor connections. Thus they might inhibit action potentials from moving along, or though excitation, encourage them upward along the line.

Some incoming "afferent" neuron axons look for a bypass. They might search for the route with the fewest interchanges, not unlike what our smartphone apps are now capable of. For example they might connect directly in the dorsal horn of the spinal cord to a different type of new neuron. This type, called a "projection" neuron sends an axon to cross the spinal cord and ascends toward the brain. After this crossover, their axons travel straight upward in bundles en route without interruption. The bundles are also called "tracts" and they extend the length of the spinal cord. Overall, the spinal cord white matter is composed of several kinds of "tracts" composed of axons going either up or down. Sensory tracts from projection neurons can include pain and temperature sensitive axons. These are called the spinothalamic tracts because they travel from the spinal cord to the thalamus in the brain. Other tracts, positioned in another area of spinal cord white matter, shuttle touch sensation up to the brain. In summary, our sensory nerve impulse journey starts from the skin and body along sensory axons in nerves, bypassing DRGs, entering the spinal cord, connecting through interneuron interchanges and projection neurons. These then form the tracts that travel to the thalamus of the brain.

Up to this point we have described how tracts in the spinal cord transmit information up to the brain. Their destination, the thalamus, lying deep within the hemispheres of the brain is also a highly complex relay station. Moreover there are additional neuroanatomical twists to this story that seem designed to generate fear in medical students learning about them. We briefly mentioned that axons must cross to the opposite side of the spinal cord en route to travel or project upward, The crossover for pain and temperature sensing axons happens at the level where afferents enter the cord, through the central zone of the cord to the opposite side, then upward through the spinothalamic tracts. In this way, while we feel pain, heat or cold on one side of our body, the tracts or pathways that transmit this information actually ascend on the opposite side of the spinal cord. The precise location where these tracts switch sides is different for each tract. You may be thinking it makes for complex wiring diagram and you are correct! However its complexity is trivial compared to what the nervous system as a whole is tasked with!

We described that some afferent axons connecting to the spinal cord must somehow bypass the interneuron traffic lights. An easier route exits.

Axons that transmit information about light touch, position or vibration have a simpler pathway. To sense the position of your limbs, sensory axons with receptors in muscles and tendons are activated. They inform the brain about what our limbs are doing in space. Without exquisite knowledge of where limbs, joints and digits actually are, it would be impossible to carry out tasks, or to walk! Instead of synapsing with interneurons in the dorsal horn of the spinal cord, these axons instead form up, without a synapse intersection, into special tracts in the very back, or posterior region of the spinal cord. These tracts are called dorsal columns, somewhat famous because they are easily seen as long bumps or smooth ridges running the length of the cord and visibly obvious without dissection. Neurologists like to talk about "dorsal column sensation" meaning sensation to vibration, position or fine discriminative touch. In the dorsal columns, the axons from the sensory neurons travel directly, without interruption, all the way up to brainstem where they finally synapse, deciding only then to connect with other neurons. Only at this level, just after that synapse, do they then cross to the opposite side. This is different than their pain and temperature wiring counterparts that have already crossed lower down. Why so varied? I cannot answer except to say that how these pathways develop during formation of the nervous system is another topic of its own. Perhaps these growing axons are attracted or repulsed along their pathways so that they limit themselves only to certain tracts, like drivers who learn to prefer some roads to others.

Thus the spinothalamic tracts, the relay stations for some, but not all sensory information, and the dorsal columns, the highways for other sensory axons, make for a complex state of affairs. This becomes evident in neurological disorders. For example in persons with severe deficiencies in Vitamin B12 or Vitamin E, there is selective damage to the axons in both of the dorsal columns. Patients experience clumsiness, loss of balance and difficulty walking. They are unable to sense vibration or position. It's a curiously localized form of degeneration in the spinal cord and when it follows B12 deficiency it is called "subacute combined degeneration." It is also a good example that the pattern of damage from many types of neurological disease, once "localized" can make the diagnosis obvious. Depending on where the problem is, some spinothalamic sensation may be interrupted on the opposite side of the body. If the dorsal columns in the spinal cord are involved, abnormal sensation is on the same side of the injury! Recall that this is because the dorsal column signals do not cross over until they reach higher levels into the brainstem. Localizing spinal cord abnormalities has been a major part of the job description for neurologists long before there were MRIs. Despite their incredible importance, sometimes MRI reports are not quite right or the images not sensitive enough, so a classically trained neurologist should be able, in many instances, to define exactly where the lesion is. Patients with an MS plaque (a site of MS inflammation) or spinal cord tumor may have loss of pain and temperature sensation on the side of the body opposite to the side of the plaque or tumor. At the same time touch and vibration may be involved on the same side.

There are engaging aspects of clinical localization. The spinal cord has a small "central" canal that contains cerebrospinal fluid. In a rare condition known as syringomyelia, this canal expands or extends into a sizable cavity, so that the pathways in the spinal cord are compressed from inside out. This differs from other problems like compressive tumors that press onto the cord from the outside. When the spinal canal fluid channel expands from within it is called a "syrinx" hence the term "syringomyelia" a cavity in the spinal cord (myelia). This central cord lesion results in a curious pattern of sensory loss in a patient. As might be expected from above, it may disrupt the sensations of pain and temperature, since these fibers cross over within the center of the spinal cord to form spinothalamic projection tracts, close to the central canal. If you have an enlarging syrinx in your cervical spinal cord, sensation over your body is altered in a "cape like" pattern that involves the skin of your neck and upper torso. Moreover, sensation to pinprick and temperature are involved whereas fine touch is preserved. The fibers that normally supply sensation to the shoulders, upper trunk and arms are interrupted as they cross, as expected, through the spinal cord normally just in front of the now expanded central canal. Syringomyelia, as one also expects, can be very painful, because of it irritates pain axons that do not like to be disturbed.

In Canadian neurological lore is the story of the late Dr. Henry JM Barnett (HJMB) an iconic neurologist who was interested in syringomyelia. Among his many accomplishments, including the discovery that aspirin prevents strokes, he was responsible for some of the very first clinical trials in stroke neurology. Barnett was an imposing mentor to many Canadian neurology trainees, including myself, and instilled the importance of classical neurological techniques, heavily emphasized in some neurological teaching but sadly often lost elsewhere. I was privileged to work with him before he retired. In addition to his work on stroke, Barnett wrote a classical book on syringomyelia.[8] He included detailed maps outlining the loss of sensation to pinprick and temperature on the bodies of his patients that had syringomyelia. Unbeknownst to the resident neurology trainees, some of these patients would reappear in hospital under his supervision for reassessment. Neither they nor we had any inkling that their sensory loss had already been mapped out and recorded by Barnett. In classical Neurology wards, trainees were required examine and present their synopsis of their neurological examination of a newly admitted patient to the chief. Barnett would listen, sometimes to long and rambling summaries by the trainees. When they were done, he would pull out his book from his briefcase and then check whether the resident was accurate in carefully mapping the area of sensory loss from the syrinx. The outcomes were that the mapping was accurate (a pass) or a sense that the presentation was selling him a bill of goods (a fail). It was best to have done a careful job before presenting your findings! To do so sometime meant staying in the hospital and spending 2–3 h reviewing the case the night before! The final point of this story is that individual syringomyelia patients, depending on the extent of their syrinx, had a very unique pattern of sensory loss on their body. Careful examination and documentation of its extent allowed clinicians to consider whether surgical drainage was warranted to reduce further damage from the cavity.

## Nodes of excitability

I explained earlier that axons and electrical wires are not exactly alike. While this is a reasonable way to begin to explain their biology, the story is ever more complicated but nonetheless elegant. Like electrical wiring, there are varying types of axons of different diameters. Unlike wires, axons are biological "soft" units composed of a lipid coating that forms a covering membrane around internal material called axoplasm. Remember that lipids are fat molecules that repel water and that lipid membranes surround all cells. The lipid membrane that surrounds the axon may be more akin to the wall of a thin water hose, although it is not a uniform material. Instead, at regular intervals, an axon has proteins inserted along it. These are called transmembrane (crossing through the membrane) proteins and they come in variety of types and functions. Many have complex folds called transmembrane domains that cross back and forth across the membrane.

What is the purpose of placing proteins within the axon lipid membrane? Everything. Without membrane proteins, axons are unable to exchange the ions that make them electrically excitable and conductors of impulses. Lipid membranes repel ions; at rest this prevents ion exchange or movement in or out of an axon. As will be discussed later, since concentrations of sodium and potassium positive ions (cations) differ between the inside and outside of an axon, an "electrochemical gradient" is created, negative within the resting axon. To set up an impulse, a circuit through the membrane is required and specific proteins called ion channels accomplish this. With appropriate prodding, a subtle change, or activation, of the sodium ion channel allows positively charged sodium ions to enter the axoplasm beneath the membrane. These cation movements, or currents, make the axon more positively charged or "excitable" and form the basis of information signaling. They can support currents that flow along the nonmyelinated axons more slowly as part of continuous conduction or more quickly using nodes of myelinated axons. Moreover, currents move by propagating through membrane ion channels all along the length of the axon, enabling long distance transmission. Slightly later, when positively charged potassium ions use their channels to leave the axon, the axoplasm beneath the membrane then becomes less positive and excitability is dampened. Sodium and potassium ion channels thus offer distinct behaviors and are designed differently. The sodium channel is structured to undergo activation when it detects positive current from a neighbor channel, an arrangement that sets up a chain of opened channels all along the axon. This positive electrical signal, or action potential (AP), thereby travels along the axon. Later, when potassium ions flow out of the axon, they reset the axon for receiving its next volley of signals. In this way, discrete packets of information, APs, are transmitted. If ion channels are defective, there may be serious neurological problems.

In myelinated fibers, nature evolved an interesting way to rapidly upgrade the speed of impulse, AP movement or propagation, alluded to earlier. Larger wires or axons transmit more rapidly by carrying more current. However, in myelinated fibers the crucial ion channels are positioned along "Nodes of Ranvier" that we discussed above. In these fibers, action potentials engage in what is called "saltatory" conduction, from the Latin "saltare" which means to leap or jump. In saltatory conduction of myelinated axons, action potentials are described as jumping from node to node. This is an oversimplification, and Campenot[9] reminds us that it is the electrical circuit that is enabled from node to node, sustaining an AP, rather than completely renewing it. The speed of transmission is, nonetheless, dramatically enhanced by 30 times or more, yet the energy cost is minimized. If large axons all had to continuously conduct, the high cost of resetting the axons for each new packet of information would be unacceptable. A series of additional proteins also populate Nodes of Ranvier. Several anchor ion channels and form connections between the myelin sheath and axon to maintain a unique structure. Nodes also have neighbors, such as the "paranode" around the node and the "juxtaparanode" around the paranode! What makes the overall node structure so interesting, unique and important is that they represent interruptions in the myelin sheath. In yet another analogy, myelin segments could also be thought of as the individual segments of cement that form a sidewalk. In between each segment are found the nodes. These might then be like the "cracks" in the cement sidewalk. However unlike simple spaces, there is a complex plan at each node to anchor SCs cells over the axon so that it and its myelin sheath remain attached. Without a way to anchor myelin, after some time it might come apart like a poorly constructed sidewalk. Poorly constructed sidewalks might not be very nice to walk along!

The exquisite anatomy of axons, SCs, nodes, and ion channels are only dealt with briefly and simply here. For a detailed and worthwhile deeper immersion into the topic several excellent references are available.[4,10–12]

## Biology bites

The complex structures of nodes, paranodes and juxtaparanodes are essential. Their detailed protein structure has been reviewed by Steven Scherer.[13,14] For example, if ion channels are not fixed within nodes, they fail to conduct impulses properly. By clustering, channels work together to mount enough current for action potentials, to jump from node to node. Of course without action potentials that propagate, sensory information cannot reach the brain and no signals are sent from the brain to trigger muscle movement! Toward this point, recently discovered neurological disorders with damage to nodes can have devastating neurological consequences. Some are genetic disorders, such as those described in a family of inherited nerve disorders known as CMTs. More about this will come later. Others are newly discovered autoimmune conditions, with inappropriate antibodies that damage node proteins. These "nodal" disorders have been discovered to cause rare forms of CIDP (chronic inflammatory demyelinating polyneuropathy), a form of peripheral neuropathy. For example, in my own practice I have followed two patients with highly unusual antibodies that damage their nodal structural proteins. These are called antineurofascin-155 in one person and anticontactin in the other. Both are very brave patients dealing with very difficult neurological problems. CIDP will also be discussed in more detail later.

## Sockets and interfaces: Nerve terminals

While we have discussed axons, equally important is how they plug into muscles and organs. These connections are varied, intricate and put our wiring analogy to shame. Unlike coaxial cable connectors, two or three pronged electrical plug ins, USB or lightning cable connectors, the biological counterparts dwarf our engineering examples by their complexity.

## Wired for muscle movement

Let us begin with motor axons. Their "sockets" or terminals determine how motor neurons that convey signals from the brain connect to muscles. This is not a simple case of attaching a cable to a muscle fiber. The terminal or end part of the motor neuron and the muscle connect to each other through a highly specialized structure, the neuromuscular junction (NMJ). The NMJs resemble the synapses of the brain that connect neuron to neuron. In fact, early anatomists and physiologists were able to probe NMJs to learn fundamentals of how all synapses might work. The NMJs are distinct from CNS synapses. An obvious difference is that a neuron, or nerve cell, connects to a muscle, and not another nerve cell. The motor axon attaches to a unique portion of the muscle fiber known as the endplate, an essential receptacle. As it turns out, NMJs involve a very close connection of terminal motor axons to endplates, the two separated by a narrow "NMJ cleft." Since electrical impulses cannot normally jump from cell to cell (we will discuss an exception during disease!), the signal from the nerve must somehow stimulate the muscle. As with CNS synapses, this is a form of short term chemical communication using substances called neurotransmitters. In brief, an electrical signal from the motor axon, an action potential, travels down to its terminal and causes calcium to enter it. Calcium within the motor axon terminal, in turn causes a large group of tiny membrane bound sacs within the terminal of the axon, known as vesicles, to empty their contents into the NMJ cleft. Vesicles in the nerve terminal contain neurotransmitters, poised for release, which are essential for communication across the NMJ cleft. For their release, vesicles exposed to a rise in calcium in the motor nerve terminal are prompted to attach to the inner side of the terminal membrane, then extrude, or dump out all of their contents out into the NMJ cleft. Of course without an action potential, calcium signals or vesicle emptying, the signal from the motor neuron to the muscle stalls. In the case of the NMJ, the neurotransmitter within vesicles is acetylcholine (Ach). Ach is released, makes its way across the NMJ cleft from nerve terminal to the muscle endplate where it attaches to and activates special proteins, the acetylcholine receptors (AchRs). For any one of the thousands of proteins, peptides and other substances like Ach made by cells, a specialized receptor is required for signaling. AchRs are indeed highly specialized.

One of the beautiful complexities of the NMJ system is that the muscle endplates on the receiving end also have a unique microstructure of their own. These cannot really be appreciated under the light microscope but can be observed when special biopsies of the muscle endplate are taken and images made using EM. The ultrastructure of motor nerve that is separated by the narrow NMJ cleft from the endplate emerges in full view. The motor nerve terminal is, as you might expect, is recognized by its content of its many small rounded vesicles containing Ach next to its membrane. At the other end of the cleft, the muscle endplate is rolled into small folded "hills" separated by "valleys." The AchRs in turn populate the hills, poised to interact as first responders to any Ach released into the cleft (Fig. 14).

Physiologists in past decades were enamored of NMJs because recordings of very tiny but random spontaneous currents in the muscle endplate could be made. These small currents arise because packets, or collections of Ach molecules are spontaneously released from the motor terminal and activate AchRs on the "hills" of the endplate. Recall that the motor terminal has vesicles packed with Ach poised for release. Each vesicle contains a packet, more properly termed a "quanta" of Ach. Since quanta have a fixed amount of 4000–5000 Ach molecules the vesicles assume a uniform size. EM images confirm identically sized clusters of vesicles in the motor axon terminal like clusters of grapes. As the contents of a single Ach vesicle or quanta is released and crosses the cleft it activates its AchR receptors and opens ion channels to allow sodium into the muscle fiber. The muscle then

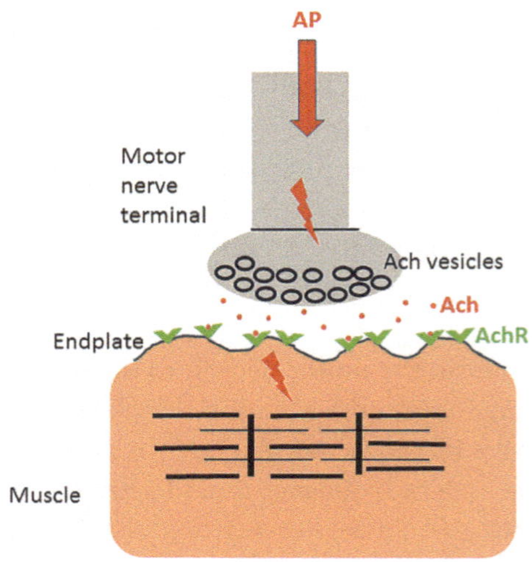


**Fig. 14** Simplified scheme illustrating the neuromuscular junction (NMJ) with the motor terminal containing Ach vesicles, Ach release and AchR (Acetylcholine receptors) on the endplate of the muscle.

is electrically activated and has its own action potentials. If enough Ach is released, the muscles are activated and they contract. One quanta, however, is not enough to do the job.

At rest, there are no action potentials and there is no flood of Ach vesicles releasing their contents into the NMJ. The junctional cleft is largely quiescent. Despite this, an occasional vesicle and its quanta is released without prompting, in turn activating one or more AchRs. The discharge from this spontaneous activation is then recorded by physiologists as small blips like backroom noise, as individual random electrical spikes. These blips, spikes or currents are officially called miniature endplate potentials (MEPPs) and their size and frequency have been used by physiologists to learn a great deal about the NMJ. For example, a reduction in MEPP size suggests something is awry with transmission. This could be because each quanta has less Ach than it should or that there are fewer AchRs at the endplate to receive them. When the blips occur less frequently than usual, the motor axon may be damaged so that it cannot release vesicles. Normal motor axons transmitting an AP do not release all vesicles or quanta into the NMJ; a substantial reserve supply of Ach in vesicles is required for followup action potentials. For example, it is common for the motor axon to fire repeatedly, in a so called "train" of action potentials. Without a reserve, vesicles would soon be depleted terminating further transmission. Muscle paralysis would ensue. Fortunately however, even with long trains of repetitive motor axon discharges, the Ach reserve is rarely depleted. However, this problem can occur in some disorders, discussed next.

## Biology bites

Why is the NMJ important? Several neurological problems target the NMJ exclusively. They are known as "myasthenic," meaning "muscle fatigue," syndromes. Myasthenia Gravis, or MG, is the most common. The main symptom of MG is muscle fatigue, which is what the term "thenia" refers to in myasthenia. "My" refers to muscle. MG is called "gravis" because it is "grave" and can lead to paralysis, inability to breathe and death. Is this convincing you of the importance of the PNS and its anatomy? NMJs are also the "socket" connections between the intercostal nerves and their muscles and between and the phrenic nerves and the diaphragm they innervate. These nerves and muscles control breathing, or respiration. MG is an autoimmune condition with inappropriate antibodies that target the NMJ, usually the AchR protein. In MG, Ach is released from the nerve but its receptors on the endplate "hills" are blocked or degraded by this auto-antibody (an antibody made against oneself) known as the antiAchR antibody. Its presence can be detected by a blood test. Thus, after an action potential signal from a motor nerve, some initial Ach quanta might activate any remaining functional AchRs, but the next signal and the next fails because what Ach is available is not enough to overcome the receptor blockade. In other words, the reserve supply of Ach, normally more than enough to do its job, is incapable of activating enough of the remaining and operating AchRs to generate a signal in the muscle. Patients develop weak eye muscles causing them to experience double vision, difficulty swallowing, weakness in their arms and legs and in severe "gravis" cases, an inability to breathe.

Why are the eye muscles targeted? This has been a topic of interest for many years. However, it may help if you consider how active eye muscles are during normal vision. Their actions are essential in making us aware of all of the visual cues confronting us and they allow us to respond. Eye muscles are continuously activated, whether you are reading, or even if you are sleeping! How does double vision, also known as diplopia, develop? In addition to moving quickly and continuously, eye muscles must work perfectly aligned with their counterparts on the other eye to point toward images. Once NMJ transmission becomes precarious, some muscles might work, but others, with AchR blockade, may fail. Failure, as in many human affairs, is an individual tragedy, targeting one muscle or muscle fiber at a time. Thus, the right eye muscles may be working normally, but if the left eye muscles have experienced fatigue, diplopia ensues. How can one be certain that it is the eye muscle imbalance causes diplopia in MG? Try covering one eye and the double vision disappears. Both eyes require exact alignment to prevent double vision.

Eyelid muscles also fatigue in MG! When we are awake, most of us keep our lids open so that we can see. Given this, lid muscles are active for most of the day and are susceptible to fatigue. When those of us without MG normally get tired our lids droop because we lose interest and fail to devote enough attention to activate the motor axons connected to our lid muscles. We then might fall asleep! Hopefully you are not doing this as you read this book! As you might then expect, people with MG experience droopy eyelids (also called ptosis) that worsens through the day. Ptosis might be complete, with closure of both of your eyelids. In mild MG, ptosis may be the only symptom, perhaps in one eye only. However this may be followed by more serious symptoms we have discussed above. Once swallowing and breathing muscles are fatigued from MG, its time for an admission to hospital, possibly even to an intensive care unit. Fortunately

MG is curable. One treatment approach is to make the Ach at the NMJ last longer. The good news is that this can be accomplished because Ach is normally rapidly degraded by enzymes in the NMJ called acetylcholinesterases. These can be inhibited by drugs, called acetylcholinesterase inhibitors, that prolong the action of Ach. The classical story of MG is a person who comes into the clinic with both eyes severely drooping, gets an injection of a drug call edrophonium, or tensilon, and the eyes magically pop open! Tensilon is a short acting acetylcholinesterase inhibitor and this test appropriately is called the "Tensilon test" for MG. By the time the person is ready to leave the clinic the Tensilon wears off and the ptosis returns. More helpful are longer acting forms of acetylcholinesterase inhibitor that can be taken as a tablet. A commonly used form is pyridostigmine, or "mestinon." Of course while acetylcholinesterase inhibitors inhibit the degradation of Ach and temporarily improve the symptoms of MG, they do not cure the autoimmune cause. Drugs that target the immune system are often required to eliminate the antibody that causes MG.

Other interesting myasthenic syndromes are inherited or are associated with cancer. All share the same issue with MG, a failure of the NMJ to properly excite and activate the muscle. Some of these disorders disrupt the nerve terminal, others the endplate. In addition, some anesthetic drugs are deliberately designed to disrupt the NMJ. The idea here is to ensure that the patient is truly paralyzed and does not jump off the operating table during delicate surgery!

## Infiltrated for sensation

Sensory axons add a completely new perspective and biology. They differ from motor axons in that they come in a greater variety of sizes and purposes. Sensory axons transmit several types of information and to do so, are equipped with unique receptacles or terminals. The first type of sensory "terminal" is simply a "bare axon terminal" without anything attached or exciting to be seen. Also called free nerve endings, these axon branches grow like branches of a tree into the outer layer of the skin known as the epidermis. Some free nerve endings travel as far as the stratum granulosum, the second most outermost layer of the epidermis. These more adventurous axons may arise from different sensory neurons called Mrgprs (Mas-related G protein coupled receptors) or IB4 (named after a type of surface protein they exhibit).[15] Other sensory free nerve endings may terminate more conservatively, in the stratum spinosum of the epidermis, one layer deeper. These are called peptidergic because they contain neuropeptides. In any event, all of these types of axons branch and separate out between the skin cells (keratinocytes) as if poised for signals. Keratin is a protein that gives structure to the skin and accumulates from dead keratinocytes before being shed as dandruff! The keratin layer offers protection from loss of water and injuries. The cosmetic industry spends billions of dollars to somehow soften or "rejuvenate" this dead outer layer so you look better. Peripheral neurobiology science might make considerable progress with 0.01% of the funding that goes into rejuvenating skin creams!

The physiology of "bare" axon function in the epidermis is not well worked out. "Bare" endings are thought to transmit pain, itching, heat and cold. Small and difficult to study, the types of proteins at their tips determines their role, specifically TRP channels we will discuss below. Stay tuned because these channels "pack a punch" if they are activated by some stimuli like capsaicin in hot foods. Alternatively other TRPs are activated by cold. Thus, with a few changes in amino acids (recall that proteins are made of chains of amino acids) and a resulting shift in the structure of the protein, different TRP channels in axon terminals relay very specific and differing information. This is in spite of the axons having an identical appearance!

Yet other sensory axons enter the skin but they provide information of another sort. These axons connect to special sensory "organs" in the skin that have a variety of names: Pacinian corpuscles, Meissner's corpuscles, Krause endbulbs, Merkel's disk. In essence these "organs" convert information about the kind of skin stimulus, to the nerve and we will discuss their physiology in detail a little later (see the review by Munger and Ide[16]). Specifically these detect mechanical sensations such as vibration, light touch or displacement.

Other specialized sensory "organs" in the skin are adapted to detect superficial or deep sensations of pressure on the skin. Activating several of these receptors, for example by stroking along the skin, sends signals up the nerve, through the spinal cord to the brain. A stroking sensation is detected. Hair follicles are also richly supplied by nerves such that when the hairs are bent or moved, a sensation is detected. The most extreme example of this response is seen in rodents. Their whiskers or vibrissae are highly sensitive to subtle movements, allowing them to navigate in small dark spaces. Thus, some organs detect small areas of skin movement while others placed more deeply in the skin detect larger areas of movement. Some respond to the speed or velocity of movement, while others just record the degree or extent of movement. Some only record the fact that movement has occurred. Overall, by using its unique repertoire of sensors, the skin learns a great deal about an object touching it-its temperature, size, movement, texture. Muscles and in tendons also have specialized sensory fibers that are sensitive to stretch. A combination of signals tells the digit or limb its location in space. This then relies on sensory fibers detecting not only muscle stretch, but also tendon tension, and joint displacement. Dermatologist Monty Lyman provides one of the better illustrations of how our repertoire of sensory sensors in the hands operate in his recent book.[17] Imagine fumbling for your keys in your pocket coming home in the dark. The cooler temperature of metal, the texture and size of the key, its movement in response to touch and the sensation of whether you have grasped it properly to unlock the doors are coded by an ensemble of sensory axons and their receptors. If not functioning properly you might be left out in the cold!

## Biology bites

Why is all of this important? The term "neuropathy" simply means a disorder of nerves-"neur" referring to nerve and "pathy" referring to damage, or disease. Polyneuropathies are neuropathies of many nerves, an overall disorder of nerves. Sensory polyneuropathies may damage some or all sensory axons, leaving other axons intact. "Small fiber" neuropathies damage only the axons that provide information about pain and temperature. Larger axons and other axons are not damaged. Paradoxically, when the small fibers are partly damaged, they can generate a severe pain syndrome known as neuropathic pain. HIV infection causes a small fiber neuropathy and neuropathic pain. In contrast, "large fiber" neuropathies target the large myelinated axons and they diminish position and touch sensation. Neurologists will notice that vibration from a tuning fork cannot be felt. Persons with this problem can be unsteady on their feet and have difficulty making fine movements. They cannot tell how their limbs and fingers should be placed! An example of a "large fiber" neuropathy is CIDP (chronic inflammatory demyelinating polyneuropathy), an autoimmune condition. In this polyneuropathy, "demyelinating" means there is damage to myelin due to inflammation. Since larger sensory axons are normally myelinated, disrupting their myelin can cause loss of touch sensation and an unsteady gait. CIDP also impacts myelinated motor axons causing weakness.

## Neglected autonomic axons

Autonomic fibers are "automatic," sending out signals to control essential body functions. Since they operate in the background, they may be taken for granted or sometimes ignored. Their routing is a little different than motor or sensory fibers and they have their own ganglia. Two different types are present: sympathetic and parasympathetic fibers.

The cell bodies of sympathetic axons are found in ganglia alongside of the spinal cord. Here the sympathetic neurons receive inputs from the hypothalamus, a deep central portion of the brain. These signals descend through the brainstem and spinal cord before sending branches through the ventral roots into the sympathetic ganglia. Recall that the ventral roots also are comprised of motor axons exiting from the spinal cord. Sympathetic fibers are activated in times of anxiety and distress and are thought of as the "stress wires" of the body. Think of the symptoms of "flight or fright" and you will recall what sympathetic autonomic fibers do. They send signals to increase heart rate, blood pressure and to cause sweating. They inhibit bowel and bladder emptying. They make your hair stand "on end" when you are frightened and they dilate the pupils. Some of their signals might not involve what you might consider as stress. They send the signals for sexual ejaculation in men during periods of sexual arousal. Sympathetic axons often travel together in nerve bundles with sensory and motor axons and are nonmyelinated. As you might expect from their role, sympathetic fibers connect to many targets including blood vessels, skin, bowel, bladder, iris muscles (that control pupil size), sexual organs and sweat glands. Despite these important roles however, they do not have specialized terminals and their axons just appear to spread out along the organ they connect to and are unstructured. Like synapses or NMJs, they send signals by chemical transmitters and most sympathetic axons excluding sweat glands, use norepinephrine (NE) as their neurotransmitter. NE is thus a stress hormone, operating alongside of epinephrine (also called adrenaline) released from the adrenal gland. Measurements of NE in the blood indicate how active the sympathetic nerves are; higher blood levels occur when the sympathetic fibers are more active. Just to be different, sympathetic axons connecting to sweat glands use Ach instead of NE.

While the role of the neurotransmitter Ach was discussed earlier with NMJs, it accomplishes very different goals at sweat glands. How? We discussed that the receptors for any signal between cells are very specific. In the case of Ach, the answer lies in their receptors, AchRs, tasked with changing a chemical message into an electrical one. Thus, some types of AchRs act on muscles at the NMJ and others induce sweating. The issue is important because cross activation would make life complicated! In MG for example, sweating is not involved in the disorder because the autoimmune antibodies do not recognize or damage the AchRs found in sweat glands. Ach receptors in muscle are called nicotinic receptors whereas those in sweat glands are called muscarinic. These names arise from old pharmacological studies of paralytic poisons called nicotine, an alkaloid from nightshade plants and muscarine, found in poisonous mushrooms. To make the biology yet more complex, we now know that there are five different muscarinic AchR subtypes, M1–M5. Nicotine has several subtypes as well because the full protein receptor has five subunits, each of which has yet further subtypes, all of which can be combined in different ways. Nicotine is found in tobacco plants and inhaled by smokers. It is very addictive, but this has probably nothing to do with its receptors on muscles. I am not sure if very heavy smokers get "twitchy" from too much nicotine but I doubt it. Their nicotine gives them a high from its action in the brain, a different story. Why you would want to inhale nicotine, knowing this toxic and poisonous origin, is mysterious!

The second type of autonomic nerve is a parasympathetic fiber. Parasympathetic cell bodies are found in ganglia associated with some cranial nerves and in nerves that emerge from the low sacral part of the spine. They connect to the pupils, salivary glands, blood vessels, the gastrointestinal tract, the bladder and sexual organs. These axons are also nonmyelinated and their roles often oppose sympathetic axons. For any organ connected to sympathetic fibers it's likely that they will also be connected to parasympathetic fibers in opposition. Thus, they slow the heart rate, lower the blood pressure, constrict the pupil and promote bowel and bladder contractions. In males they promote erection as opposed to ejaculation. Like sympathetic fibers, parasympathetic axons also communicate to organs with a chemical transmitter. This is also Ach! Ach from parasympathetic fibers activates muscarinic AchRs.

## Biology bites

Autonomic axons are frequently damaged in neuropathies. For example, persons with neuropathy may lose their ability to sweat in their feet. When autonomic axons that control blood pressure are damaged, simply standing up may cause a drop in blood pressure and fainting! This is because standing normally activates sympathetic nerves. The sympathetic nerves send signals to blood vessels throughout the body telling them to constrict and counteract any fall in blood pressure that may result from standing. These blood vessels are also called "resistance" vessels. Their walls contain "smooth" muscle fibers that contract from a sympathetic signal, causing constriction. Veins have fewer smooth muscles and are less able to constrict. Smooth muscle fibers are not the "skeletal" muscle fibers found in your arms and legs. Many years ago, anatomists compared muscles in the arms and legs with muscles controlled by autonomic nerves. Despite both being muscles, they looked different. Skeletal muscles are striped or "striated" from their well organized contractile proteins within whereas smooth muscles are not because their proteins are not as well organized. Since smooth muscle fibers are activated by sympathetic or parasympathetic axons, you do not have voluntary control over them. Thus, the iris of the pupil, the constriction of airways, the contractions of gut and the spasms of a full bladder all depend on smooth muscles connected to autonomic axons. NE is the critical neurotransmitter then that activates the smooth muscles to narrow or constrict resistance blood vessels. Ach acts on sweat glands, salivary glands and other targets of ANS axons. This becomes important because some drugs, for example certain antidepressants, have the unfortunate side effect of inhibiting how Ach acts on its receptors. Persons on these drugs might experience a dry mouth, constipation or a bladder that will not empty.

Many people, especially the elderly, experience postural hypotension, a significant enough fall in blood pressure upon standing that they feel faint. Serious consequences follow fainting and falling including fractures and head injuries. Older persons may develop postural hypotension because they have impaired sympathetic nerve function. Some disorders, such as Parkinson's disease, a CNS problem, interrupt sympathetic responses and cause postural hypotension. Most commonly however, medications designed to control high blood pressure have the side effect of blocking sympathetic signals. If so, the type and dose of the antihypertensive medication may need to be reconsidered.

An early symptom of autonomic damage is erectile dysfunction in men. This is common in diabetes mellitus and other disorders that damage autonomic nerves. Normally parasympathetic axons, with some help from sympathetic axons, dilate blood vessels that allow engorgement of the penis and erection. Ejaculation, in turn is signaled strictly by sympathetic axons. These connect to related organs including the epididymis, vas deferens, seminal vesicles and prostate gland that are involved in secretions and moving spermatozoa along. A disorder of the autonomic nerves, or use of a drug that disrupts the nerve action can cause erectile dysfunction, or ED as it is known by its television advertisements.

Some persons develop widespread damage to their autonomic nervous system that has not been recognized. This is because relatively few laboratories are capable of administering the specialized tests required to diagnose autonomic disorders. Phillip Low, a neurologist and a personal mentor at the Mayo Clinic, has been a pioneer in quantitative evaluation of the autonomic nervous system and in describing its disorders.[18]

**Note added in proof:** Abdo and colleagues have identified a previously unrecognized population of glial Schwann cells at the immediate subepidermal layer, with some limited extension into the epidermis. These have a functional role in noxious thermal and mechanical stimulation.[18a]

# CHAPTER 2

# Constantly think physiology: Structure meets function

## Remarkable conduits

Axons transmit information rapidly. An action potential, the unit of information in the PNS, travels along larger myelinated axons at 50–60 meters per second (m/s). This may not sound very fast to an engineer, particularly someone thinking about fiberoptic cables. For the human body however, this speed is remarkable. Remember that all meaningful transmitted information needs to be eventually processed by the CNS. Complex CNS or brain processing may in turn require seconds for impulses to be processed. Information traverses multiple synapses and neurons before they are placed into context and have meaning. Ideas, thoughts and awareness take time to be generated, vastly slower than the time it takes for the information to arrive by nerves. Soft biological systems are not robust enough nor wired properly to mount nanosecond reflex responses. Even if we were capable of making them, responses made too rapidly could damage muscles, joints or skin. Instead, more careful graded responsiveness that allows error or "play" is a much safer evolutionary solution.

Unmyelinated axons conduct more slowly, from 0.5 to 2 m/s. Despite this, one would hardly notice the difference. Activating sensory receptors, such as those found in the skin may take longer. Similarly NMJ transmission involves mobilizing vesicles within the motor axon ending, Ach release, Ach travel across the cleft, its attachment to receptors and its eventual activation of the muscle. The total delay is thought to less than 1 ms, but in this interval, a nerve might send an action potential a distance of 50 m! Given these facts, it seems unlikely you would be able to distinguish whether your sensation is transmitted from myelinated or unmyelinated axons. This is probably true. However we will later discuss "slow" or "fast" pain, a possible exception to this idea. To summarize, while some axons transmit over tenfold faster than others, more significant delays involve receptor activation and CNS processing.

*Our Wired Nerves*
https://doi.org/10.1016/B978-0-12-821487-9.00002-7

Despite the debate over whether faster conduction would be useful to human organisms, other aspects of their function are impressive. Axons can conduct "trains" of signals rapidly, from single action potentials, to repeating ones at 200 times per second (200 Hz). This is enough to recognize that a neurologist's tuning fork is vibrating at 128 Hz. Enabling such information transmission involves a remarkable molecular arrangement that refreshes the nerve membrane or node for the next impulse. This will be also discussed below.

## Action potentials: How the connectome is activated

How does the connectome activate? Axons in nerves send signals by electrical impulses called action potentials (APs). They signal action and are named "potential" in line with engineering ideas around electrical potentials, also called voltages. Actual voltages in axons are not very high, but high enough to transmit information. They are not felt as a shock. A household outlet at 120 volts delivers a shock from the socket and may electrocute the unwary. Nerves operate in the millivolt (mV) range, one thousandth of a volt. We would otherwise be shocking to one another! Unlike the current coming out of a household electrical outlet, they are also not "alternating currents" or AC. APs are a type of direct current, also called DC. We may have alternative lifestyles but we do not have alternating currents. Perhaps this is why sticking a fork into an AC electrical outlet feels so unpleasant and is definitely not recommended. The DC currents in nerves have interesting properties of their own.

APs are "all or nothing". In other words they either fire off, or they do not fire. You can think of them like projectiles from artillery cannons that are either shot or not. Like ordnance from a cannon, once an AP is launched it can't be brought back. Axons then must be reset for the next AP. To be precise, activation of APs occurs in the membranes containing ion channels and the adjacent axoplasm immediately inside of the membrane (Fig. 15). Resting axons are normally charged negatively, in the range of −70 mV, known as their resting potential. Why the inside of the axon should have this electrical charge is complex, but explained by different concentrations of sodium, potassium and chloride ions inside the axon membrane than are found outside. Sodium levels are much lower inside than outside, whereas potassion concentrations are the reverse. Active pumps in the membrane, using axon energy stores, pump sodium out of the axon, in exchange for accepting potassium inside to maintain this imbalance. These differences in concentration set up an electrical gradient, or voltage from inside to out and

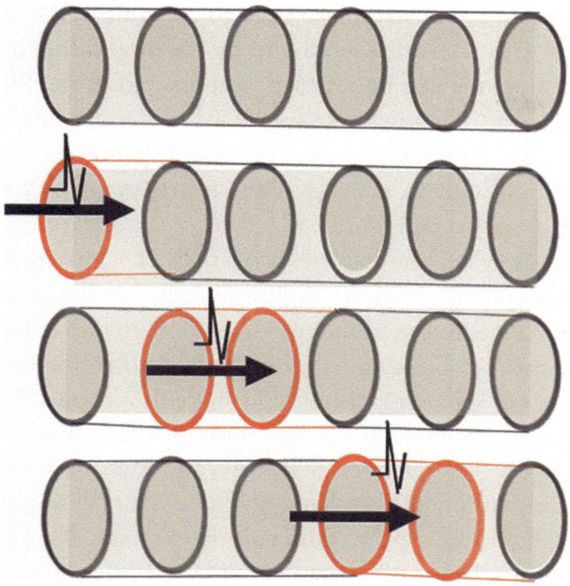

**Fig. 15** Simplified scheme showing propagation of APs (action potentials) based on excitability of the outer rim of the axon, beneath its membrane.

the membrane prevents free flowing ions from equilibrating. The overall resting potential depends on the balance of all three ions, but mainly on sodium. The explanation here is a simplification and the reader is referred to detailed explanations of this fascinating biology.[9,10] Any added electrical impulse then temporarily adds positive charge to make the axon somewhat less negative. Is it enough to generate a signal, an AP? It may not be. Axons are a little difficult to please, and require enough of a jolt, or stimulus to readily to launch an irreversible AP.

Axons filter out ongoing traffic by not responding to random stimuli. To launch a propagating AP there needs to be sufficient signal to reach a "threshold" for firing. This usually means shifting the axon to less a negative charge, called "depolarization." As this shift occurs long before it reaches 0 mV, the axon reaches a threshold for firing off our projectile, the AP. Threshold is interesting, usually thought to be approximately −55 mV and it represents an inflection point, where axons now acquire the wherewithal to generate APs. Biochemically, it means that at threshold, a series of proteins of the axon membrane, not normally responsive to small subthreshold stimuli, change their properties. They now acquire the configuration or 3D structure primed to participate in AP propagation.

"Excitability" of the axons simply refers to how close the electrical potential of an axon is to its threshold and hence how difficult it will be to activate it. Some axons can be set to be yet less excitable. The term "**hyper**polarization" means the charge in the axon is more negative than the resting membrane potential and as a result, less prone to fire APs. These axons are more difficult to activate. This shift is opposite to **de**polarization, a change closer to threshold.

What happens when the axon is activated? The ion channels embedded in the layers of the axon are highly sensitive to the electrical charge around them. When this environment changes, there are subtle changes in how these proteins are fitted together or bent in three dimensions. A little encouragement, by achieving threshold, suddenly permits a series of ion channels proteins to allow ion passage through their pores. All along the axon, boring stable ion channels suddenly are contorted a little because their electrical environment has changed. There is a rush of positive sodium ions through the channels, propagating along the axon. Think of the AP as having momentum, like our projectile. As the channels open, the next set of ion channels activate along the axon and so on down the line, bringing a host of neighboring ion channels into play. You might then imagine a "wave" of these open channels spreading all along the axon, termed "propagation" by physiologists. The actual size or voltage of an AP measures how far the electrical potential changes from the resting potential. How intense is the charge that develops? APs overshoot the threshold potential, pass 0 mV and may reach +30 mV. Remember that a car battery is 12 volts, over a thousand times more powerful. So do not expect your nerves to jump start a dead battery, particularly one frozen during an Edmonton winter.

Why doesn't the axon now stay "depolarized" or positive once an AP forms? Why is it a transient "wave"? Mother Nature has built more than one kind of protein into the membranes of axons. There are those that allow sodium ions to flow into the axon to depolarize it (less negative) and form the AP. However there are others that only allow ions to flow out of the axon, the reverse direction! When this happens, the axon loses positive ions and returns to being negative, perhaps hyperpolarized, more negative than resting. If these contrarion ion channels were to open up at the same time the AP forms, then nothing interesting at all might happen. The nerve would not become more positive or negative and the story would end there.

The beauty of AP formation is in the sequence with which protein ion channels open. When the axon is ready at threshold to fire an AP, only inward flowing sodium ions are allowed in. Sodium ion channels are the first to be activated. Not to be outdone, contrarion potassium channels next open up and negate whatever has happened a few milliseconds later. These channels allow positive potassium ions to flow out, the reverse direction. The APs are abruptly shut down! The whole process is not unlike a hallway with a series of doors. One opens, then the next and next. However, shortly after the first door opens, it gets slammed shut, followed by the next and next. The opening and closing doors move down along the hallway and the irritating noise is "propagated" down the whole hallway. Overall then, APs are transient. They discharge, move along the axon, then disappear. After they are gone, the axons needs to rapidly reset in case there are several hundred more APs on the way. By rapid resetting, APs as frequent as 200 per second or higher can move along the axon requiring ion channels to work within milliseconds. Obviously, its not only a single AP that is important in signaling, but their numbers and how frequently they come along, their firing frequency. Our contrarion potassium channels are mainly responsible for the reset, making the inside of the axon membrane negative again. There is an additional way to shut down APs beyond outward moving potassium ions. APs can be terminated when the axon is at its most positive, such as +30 mV, by sudden inactivation or closure of the sodium channel, the original culprit. It is as if sodium channels can only tolerate so much activity, bending or deforming in response to the positive charges and then suddenly closing passage to more ions. Some electrophysiologists have argued that sodium channel inactivation is more important than potassion channel activation in limiting the size of the AP. The APs stop growing, perhaps because of inactivation, begin to decline, then shortly thereafter are fully shut down by outward potassium ion protein channels. After an AP passes by, axons have a short period of hyperpolarization, known as the refractory period. This temporary loss of excitability prevents any given AP from abruptly reversing direction, despite the fact that the axons otherwise have no sense of direction.

We can stretch our analogy a little further beyond doors opening and closing all the way down the hall. The hall might be full of people but many are very grumpy and silent, perhaps hyperpolarized. Among the grumps there are a just a few somewhat more cheerful talkative people called potassium types that make the gloominess a little less bad. Suddenly a door opens and in marches some very cheery and boisterous sodium type people. Their job is to walk a little further down the hall and open the next door to let in more cheerful people. The mood of the hallway is changing for the better. However, the doors become more difficult to keep open.

Meanwhile the potassium types, not in favor of too many sodium types in the hallway begin to leave. Without more cheerful people arriving, the hallway gets a little grumpy again, back to the way it was originally. Not only have fewer cheerful sodium types stopped coming in, the somewhat more cheerful potassium types have also left in good numbers. The axon, or hallway, resumes its silence.

If an AP, once activated, sets in motion the ion channels of its neighbors on either side of its section of membrane, how do they travel in the correct direction? For sensory axons this signal needs to travel from the toe or finger to the brain. For motor axons APs require direction from the spinal cord down to the muscles. This is a very important point! However, it is the anatomy of each of these axons that determines how APs move along it. For sensory axons, activation begins in its bare endings or terminals out in the skin, discussed above. The end of the sensory axon is where the AP is set up. Once past threshold, the axon activates ion channels all the way up the sensory axon to its connections in the spinal cord. As each AP moves in this correct direction we know that they are inactivated, then quickly prepare themselves for another AP to come along. They have no opportunity to move backward because the reverse direction is now off limits. In motor axons, the signal comes from brain, to spinal cord to motor neurons in the spinal cord. You need to signal your muscles to move, an idea that originates in the brain. Once activated, the spinal motor neurons send along their APs outward to muscles, in a direction opposite to sensory axons. Motor and other CNS neurons have a specialized segment called the "axon hillock" where their cell bodies and axons connect. The hillock is built with the correct structure and content of proteins that elegantly and easily generate motor action potentials. Consider next what might happens to an axon that happens to gets activated half way along its pathway, neither near the muscle or spinal cord? Fortunately this does not happen very often, unless the nerve is abnormal, a state of affairs discussed later under "Action potentials out of control." Nerves and axons are not prone to be activated except at sensory terminal areas or in motor axon hillocks. Imagine the confusion otherwise!

There is an exception to this rule that is artificial. When a person undergoes electrophysiological testing, described in a later chapter, the rules of AP direction are deliberately disrupted, albeit only temporarily. During these studies, also called nerve conduction studies, electrodes are applied to the skin near a nerve to activate all of its axons. This requires more intense stimulation because it is designed to activate all of the axons in the nerve. You feel a shock. Fortunately most people that undergo these tests tolerate the shock well, although women characteristically tolerate them better than men! Normally it would be rare for any of us to activate all of our axons in any single nerve. I doubt if its possible. In any event, such an artificial state of affairs can be carried out in the laboratory; the stimulus generates APs where it is applied. Because of the structure of the nerve, APs then flow in both directions away from the stimulus. It flows down the nerve to activate any muscles in its pathway. It also flows up to the spinal cord and generates a sensation-a shock. With careful recordings, either event can be recorded and documented depending on whether you record further up the nerve, or down the nerve toward the muscle. There are complicated names for these scenarios. When electrophysiologists stimulate a nerve and record its APs in the same direction nature dictates, its called an orthrodromic response. An orthodromic response means the AP travels in the direction nature intended. For a motor nerve it means stimulating its axons higher up and recording further down in the muscle. For a sensory nerve it means stimulating out in the periphery and recording in the nerve closer to the shoulders or hips, higher up. "Artificial" conduction created in the laboratory can also go the opposite or wrong way and this is called "antidromic" conduction. Thus, you could stimulate sensory axons up in the arm and record responses in the fingers-the reverse direction of normal conduction. The interesting fact after all of this discussion however is that the speed of conduction in either direction is identical! Normal orthodromic, or artificial antidromic conduction in nerves moves with identical speed, known as conduction velocity. This is helpful to electrophysiologists who need to take a measurement and may not be able to set up their electrodes exactly as nature might want them to. It also speaks to the exquisite design of the system-a cascading series of electrical waves that move along the nerve during normal physiology in a way designed to transmit information but without requiring complex direction sensors. It also reinforces the idea that axons are really like hollow cables. The direction APs travel does not rely on the way the axon is structured but instead depends on where they are generated. Refractory periods following APs and mentioned earlier, are only temporary blocks of "backward" conduction and otherwise do not dictate which way axons can transmit.

Finally, what about the issue of conduction in myelinated axons compared to bare, or unmyelinated axons? The electrical theory behind conduction in axons was first described by the neuroscientists Hodgkin and Huxley. To do so they studied giant squid axons. These axons are much larger than their human counterparts and allowed electrodes to impale them to learn of their properties. Giant squid axons behave like "bare" or unmyelinated axons by sending APs along them using "continuous" conduction. Myelinated axons are an evolutionarily advanced version in which APs are not required to continuously conduct. Rather, excitability or depolarization can jump from one node of Ranvier to the next, "saltatory" conduction described in the last chapter. In between the nodes, the myelin sheaths provide insulation and protect the current to permit its extending to the next node. The cost of conduction is considerably less because only the nodes have to be activated. Conduction velocities of APs that can jump from node to node are much higher than those from unmyelinated axons (Fig. 16).

While I have oversimplified the physiology of APs here, the basic principles are correct. Once understood, the nuances of APs become quite interesting, but its not a topic to discuss at this time. Indeed, there is a remarkable collection of additional ion channels to sodium, potassium and also to calcium that we did not discuss. These come with interesting names such as Nav1.3, Nav1.7, delayed rectifier and many others. Finally, more exist and operate in the brain and spinal cord. This makes sense since control of APs in highly sensitive CNS circuits probably requires yet much more vigilance and control. For further details on axon conduction and physiology, please refer to detailed and thorough reviews of the process[10] including a recent book by Robert Campenot at the University of Alberta.[9]

Imagine a depiction of the human body color-coded to show the nervous system and all of its actively discharging nerves! While this technology is still unavailable, consider a complex recording system detecting every microscopic axon and its activity state throughout the body. Now imagine the act of walking while viewing APs in real time. We might see sensory

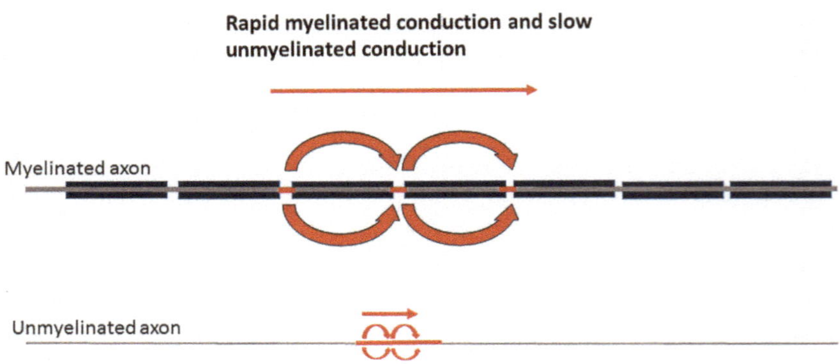


**Fig. 16** Diagram comparing saltatory conduction along Nodes of Ranvier in a myelinated axon compared to continuous conduction in an unmyelinated "C" axon.

axons in the sole of the foot activated as they touch and carry our weight along the floor. We would see sensory axons discharging in many of our joints because they bend when we walk. Sensory axons in the joints are exquisitively sensitive to movements. We would see motor axons sequentially activating all the muscles involved in walking including those that swing our arms. Motor axons even supplying the neck would be activated to keep it upright and involved in the walking motion. However, it is not just motor axons supplying muscles that would light up. Every time a muscle moves or contracts, its sensory axons are activated by the movement. Many more nerves than I have mentioned would come into play. What a marvelous image would emerge! Of course the brain would also light up, not discussed here, depending on how engaged someone was in thinking while walking. Some of us who walk into doors might arguably not be coordinating their brains with their walking pattern. Alternatively, the brain may be active, but focused on something light years away from walking! Most people can also chew gum and walk at the same time, so please pay attention to the nerve supply to jaw muscles when these images are created!

## Action potentials out of control: Fasciculations, pain and other troubles

Do APs sometimes fire inappropriately? Yes, if you consider that many people talk inappropriately, even when supposedly leading a country, this makes sense. Somewhere in the brain of someone who says something nonsensical, that idea or thought must originate as an AP in a neuron. This neuron and AP then stir up an initial circuit, activate a whole series of additional circuits and finally activate motor neurons to the muscles participating in speech. The motor neurons dutifully convey the words no matter how regrettable.

However by inappropriate here, I mean APs that appear in nerves without a command to move and no new sensation to report. Given the exquisite sensitivity of the axon membrane and its proteins poised to convey messages, a slight change might serve to trigger an AP when none was expected. A good example follows a decrease your blood calcium level, arising from an intestinal disorder or another problem. One result of low calcium is tingling in the hands and feet, a symptom that does not convey to the brain any form of actual sensation. These tingles are simply odd sensations, arising from APs that come without a direct or appropriate stimulus and they can be painful and irritating. Low levels of calcium change the excitation properties, or pore forming capabilities of axon proteins allowing axons to depolarize spontaneously. Once past threshold, an AP discharge occurs. If you have experienced tingling, you know that it is an unusual feeling, not quite like touching anything. It appears to arise from a variety of places on your body, requiring several nerves to participate and it seems difficult to localize. As is the case of normal sensation, the CNS filters out extraneous "noise," but pays attention when there are enough persistent discharges, perhaps at high frequency rates. Annoying symptoms like tingling require sufficient numbers of activated axons firing at a high enough frequency for us to notice. In the case of low calcium, axons in several nerves may be randomly involved depending on how receptive they are to unwanted depolarization. Of course, if you already have a nerve injury the axons within that nerve may be particularly prone to inappropriate excitation and the tingling might be confined to its territory. Persistent and severe sensory axon excitability can be associated with pain. Even if only a fraction of the sensory axons, such as pain transmitting axons, are overactive, abnormal and unwanted pain may arise.

What happens when motor nerves are overactive? As you might expect, they set off unwanted muscle contractions. Some are minor inconveniences. Fasciculations are unwanted and involuntary twitches of groups of muscles that are transient or sometimes persistent. Many of us have had fasciculations our eye muscles, especially the orbicularis oculi muscles that close the eye. When activated we may appear to be winking. Caffeine beveridges including "hitest" forms of coffee sold in popular coffee bistros predispose you to fasciculations. I am not aware of a systematic review of fasciculating orbicularis oculi in customers who are visiting Starbuck's or Tim Hortons (for nonCanadian readers, this coffee shop is a national symbol). When in coffee shops however, be aware that a person may not be intending to wink at you. Caffeine, added to sleep deprivation and anxiety, can generate frequent fasciculations. Medical students are sometimes concerned that they have developed a serious neurological disorder when their caffeine intake, combined with sleep deprivation and anxiety, induces persistent twitching. These are all benign fasciculations and arise from spontaneous APs in motor axons. Other than setting up inappropriate familiarity in the coffee shop, these fasciculations are not to be feared.

There are more types of abnormal muscle contractions arising from spontaneous APs in motor axons. Cramp discharges can be quite painful even if they are benign. At night they are often called "Charley horses." This seems to be an unusual label for inappropriate motor axon APs. According to one internet source (www.historybyzim.com/2013/03/charley-horse/)

a baseball pitcher named Charlie "Old Hoss" was famous for having leg cramps. You can then surmise why this name might have been modified over time to "Charley horse"! Cramps can also occur in other muscles if they are injured or damaged. A more extreme example is "tetany," set off by our example of low calcium. In tetany, the forearm muscles develop a severe cramp causing the wrist and fingers to painfully flex when a cuff is inflated around the arm to above the systolic blood pressure. In fact, this unusual symptom has been used by clinicians for many years in order to identify low levels of calcium. It is called "Trousseau's sign" or a more descriptive medical term "carpopedal" spasm. Yet another symptom of low calcium is "Chvostek's sign" in which tapping over the facial motor nerve below the ear causes the facial muscles to contract. Like tingling in sensory nerves, both are signs of hyperexcitable motor axons.

There are a few additional types of interesting movements that arise from overly excitable motor nerves. One rhythmic form, "myokymia" can develop in facial muscles where it causes a curious and repetitive movement that differs from simpler fasciculations. "Stiff person syndrome" was originally termed "stiff man syndrome" but the name was gender problematic since women are also impacted. In this condition, muscles of the arms, legs and the back contract inappropriately. The back muscles can stiffen and extend so powerfully that persons with the condition can fracture their vertebrae! Stiff person syndrome is a rare autoimmune condition caused by an autoantibody formed by the immune system against a molecule known as GAD (glutamic acid decarboxylase), that normally synthesizes a neurotransmitter, GABA, that inhibits motor neurons from firing. Since GAD is targeted, less GABA is produced and its normal inhibition of motor neurons is reduced. Whole groups of motor axons begin to fire and muscles contract inappropriately, sometimes very painfully. A person with stiff person syndrome can be disabled and it may take many medical visits to identify the correct diagnosis. The medication Valium (diazepam) is an established and effective treatment. You may recall its bad reputation as a favorite suicide drug in sad but epic stories like "Valley of the Dolls." Diazepam in high doses causes coma and impairs breathing. Originally used as a sleeping medication, its duration of action was prolonged and it became addictive. It is also used in Neurology for terminating seizures. For stiff person syndrome, the drug precisely targets the molecular changes in motor neurons that cause them to discharge inappropriately. Diazepam is a GAD agonist (agonists activate, antagonists block), thereby supporting its production of GABA and its inhibitory role.

There are a large number of other disorders, called movement disorders, that arise from unwanted neuron discharges. However these mainly arise in the brain. They include conditions such as Parkinson's disease and Huntington's disease, not considered further here. There is no shortage of work for neurologists given the wide variety of things that can go wrong with the nervous system!

## Other things move within axons: Axoplasmic transport

APs are not the only things that move along axons. Axons can be thought of as highways for a variety of items that move along more slowly than electrical discharges. These are carried in both directions along the axon, a process known as axoplasmic transport. Axons normally contain microtubules, structures that resemble a series of pipes and are essential for moving cargo. Proteins, organelles and RNA attach to the outside of the pipes and are moved along like a conveyor belt. Organelles are complex substructures within cells. Some suggest many of our organelles originally arose as donations from bacteria millions of years ago that formed a symbiotic, or interdependent relationship. The most important organelles within neurons and all cells are mitochondria. Mitochondria are essential for generating energy to keep the cells alive, allowing protein production and supporting the electrical charge of the axon. Without mitochondria, all cells including neurons die. Watching healthy mitochondria movement has become more than a pastime. By using special dyes that label and make them visible in imaging techniques, investigators have acquired important information about nerve biology and specific neuropathies. Continuous imaging of mitochondria illustrates how they move back and forth along the axon, guided by the microtubules, sometimes pausing perhaps to deliver an extra dose of ATP energy to a site in need. Mitochondria sometimes divide into daughters (fission) or fuse back together with a neighbor. Specific proteins, such as mitofusin, help to control fusion and DRP1 (Dynamin-1-like protein) regulate mitochondrial fission. This is important because mutations of mitofusin can cause inherited polyneuropathy. How fusion and fission of mitochondria become indispensable for axons and neurons is interesting to consider, but unsolved to date.

In addition to mitochondria, further organelles move along axons. Small packages called vesicles contain proteins, RNA and other contents. The vesicles we have discussed at motor axon terminals are one type. In the case of motor neurons, it is thought that the Ach necessary for transmission is made in the cell body, then shipped in vesicles to the motor axon terminal.

Like small plastic bags or balloons, they travel along the microtubule pipes as part of axoplasmic transport. Transport can be directed from the cell body out to the end of the axon, so called anterograde transport, or in the reverse direction called retrograde transport. Both directions of movement are essential. While not attaining the speeds of AP propagation, anterograde transport considered to be fast, is over 400 mm/day whereas the speed of retrograde transport is somewhat slower at 200–300 mm/day![10] Through anterograde transport, proteins, peptides and other molecules are shipped down to the end of the axons where they act as neurotransmitters. Other proteins and the lipids that form the vesicles may help keep the ends of the axons intact and in good repair.

Both anterograde (forward) and retrograde (backward) transport along axons require energy and mitochondria are readily available to be at their service. In a sense, moving mitochondria pay for their own passage. Axoplasmic transport requires two critical and interesting proteins to make it work. The first, called kinesin (like kinesiology, or the study of movement) helps to propel forward anterograde transport. Dynein, in contrast, helps to make reverse, or retrograde transport happen. Both are called "molecular motors." Why is this important? Recently yet further types of inherited neuropathies have been discovered with abnormal "motors." Without proper transport, nerves die.

If anterograde transport maintains the upkeep of axon terminals, what is the role of retrograde transport? Cargoes that move backwards to the perikarya are interesting. For example, nerve terminals are capable of molecular uptake from their local environment, packaging their contents into vesicles then shipment toward the cell body. One of the classical growth factors, NGF (nerve growth factor), is taken up and retrogradely transported by some types of sensory neurons. NGF transport is essential for survival of neurons in embryos. In adults, its uptake may not be essential for survival but instead promotes their overall health and function. In some neurons NGF causes axon sprouting. How a retrogradely transported growth factor like NGF alters overall behavior of the neuron remains to be discovered.

Some types of retrograde transport are not helpful. For example, both the rabies and herpes zoster virus (shingles virus) are transported to the CNS in peripheral axons. Researchers have also designed retrograde toxins, based on proteins called ricin, abrin, or saporins and also called "suicide" proteins. These unfortunate travelers are taken up by nerves and transported, bringing death to the neuron. Suicide proteins can be used by neuroanatomists to study and trace complex neuroanatomical connections.

While fast anterograde and retrograde transport in vesicles are appealing and are not unlike high speed trains, there are other modes of moving through the axon. Another is known as slow axonal transport and it operates at only 0.2–2.5 mm/day.[10] Why some constituents require fast transport and others slow is not really clear. However, it may be that slowly moving items are important but not time sensitive. They are like slow barges floating down a river with a load of wood or coal. Neurofilament subunits that form into the polymer scaffold of nerves, are moved along the axon by slow axonal transport.

One class of molecules was not recognized as a transport cargo because of long standing dogma. In the nuclei of neurons and other cells, DNA is transcribed into messenger RNA (mRNA). The next step involves translation, in which mRNAs are synthesized into protein by ribosomes. Dogma has stated that both transcription of DNA and translation of mRNA exclusively takes place in the perikarya, or cell body. In neurons with long axons, it was held that only protein endproducts would then be shipped by vesicles down the axon. However neuroscientists Jeffery Twiss, Diana Willis, and Christine Holt, among others, discovered that the ends of axons and growth cones can translate as well.[19,20] By this mechanism, mRNAs are carefully transported along the axons to where they are used to make protein. mRNAs can await their time of need from a site nearby, rather than a remote cell body. This alternative view of the world has helped to explain how sustaining the ends of long axons with critical proteins would be feasible. While antergrade transport is labeled "fast," it may not respond as quickly as needed. Alternatively, if local sites of the axons generated their own proteins from mRNA, we can imagine a type of flexible and responsive form of local control. The best analogy would be in identifying who might repair potholes in the road you take to work. This is an important question in Edmonton, from where I write this. How responsive to your pothole needs would the prime minister in Ottawa be compared to the mayor of Edmonton? In Alberta sometimes we see signs that state "More Alberta. Less Ottawa". I don't think however they were thinking about local mRNA translation in axons (or about potholes for that matter!).

In summary, axoplasmic transport is a process that occurs routinely in nerves. You might think about axons like roadways, railways or river routes, constantly carrying moving cargoes. New technology allows neuroscientists to see this movement, including that described of mitochondria, along axons and its very instructive! The transport vesicles, for example, speed along, stop, pause, reverse direction then move forward again. These images are different than our roadway analogy because the vesicles move so quickly and seemingly erratically that it would make for unsafe driving. Despite this bizarre state of affairs, none of this careful scientific work has discovered vesicle crashing or cargo mishaps. It all seems to work out.

## Do I have good reflexes—What are these?

When we think of reflexes, we think of how quickly we react to circumstances. A hockey player about to be checked by a hit from an opponent throws the puck over and past their stick in a flash. An outlaw draws his pistol first in a duel and wins. A biathlon athlete skis faster than all his or her opponents, then calmly lies and fires a rifle into a small target repeatedly and highly accurately.

How quickly and accurately we respond may be a type of reflex. It is often thought that these "good" reflexes can occur very rapidly without thought. However, in the context of axons and neural connections, they are slow as molasses. The reflexes we commonly consider, like those of the outlaw or biathlete, involve a sensation, signals that traverse some part of the brain followed by a motor response or action. While they might seem fast, they occur in seconds and they actually involve thought, although it seems to happen quickly. The brain is an obligatory part of the action.

Neuroscientists think about reflexes in a different way. All of the above examples are complex reflexes or reactions that involve multiple nerves, synapses and APs. Reflexes can be divided into complicated sounding names such as "monosynaptic" or "polysynaptic." Polysynaptic involves many synapses (poly) in their pathway. As previously described, a synapse is a specialized junction in the CNS that connects neurons. The complex "reflexes" we discussed in our examples above are all polysynaptic and involve brain connections. Some might not even view them as true neurobiological "reflexes" because they involve brain cortex processing.

Let's start with the simplest and most fundamental reflexes. These are monosynaptic, with only one connection to make. Neurologists spend a lot of time on these and even carry special hammers to test for them. Another name for basic monosynaptic reflexes is "deep tendon reflexes." Only about five of these are routinely tested on each side of the body, and they all involve stretching a tendon. You will recall that tendons have sensory axons and endings within them that tell us what degree of tension they are under. Like skin axons, these might be rapidly adapting, sensitive to quick movements or slowly adapting, sensitive to longer periods of pressure. A "deep tendon" reflex starts by triggering sensory nerves and endings in tendons that are rapidly adapting. They sense something that has happened quickly. So, when your favorite neurologist taps (does not bang!) your tendon with a "reflex hammer" (these hammers are not suitable for nails or carpentry!) that has a rubber tip, it bends the tendon. The best example is the quadriceps tendon just below your knee cap. The leg needs to be nicely at rest and a quick short painless tap activates the sensory endings in the quadriceps tendon. What happens next?

You may feel the reflex hammer tap, but that is not the true deep tendon reflex. If in a bad mood, or if you are surprised by the tap you might grimace and move your leg away, as if something noxious happened (it did not). This is also NOT the reflex. Neurologists become expert in recognizing whether someone is annoyed and surprised, and pulling their leg away, as opposed to having a true knee, or quadriceps deep tendon reflex. A true deep tendon reflex happens faster than you can pull your leg away and should only occur with a contraction in the correct muscle you are testing-the quadriceps! When the quadriceps contracts, your leg kicks out!

The sensory axons stimulated by the tap send signals back up to the spinal cord. Like all sensory signals, they enter the dorsal horn of the spinal cord where the single synapse takes place. The incoming sensory nerve connects through this synapse to an outgoing motor nerve within the anterior or ventral horn of the spinal cord. If the tap activates enough sensory signals, something that that does not require massive force (!), enough motor neurons send signals back down to the quadriceps leg muscle to cause a contraction. And behold, the leg kicks forward, the response expected from activating the quadriceps muscle. Taken together, it's a three part story with a single synapse. The tendon sensory axons are triggered, their APs travel up and synapse in the spinal cord and motor neurons are then stimulated to send AP signals back down to the muscle. You definitely do not need to think about this for it to occur; the cortex is not at all required. Decerebrated (brain removed) animals, that no longer have a connection of their brain to the spinal cord, have perfectly functioning deep tendon reflexes. This is not to say that the brain and upper spinal cord do not have any influence over reflexes. Their impact involves "tuning" them up or down. Anxious persons can have quite brisk deep tendon reflexes as new medical students discover when they test their new hammers. As it turns out, anything that disrupts the normal descending motor pathways from the brain to the spinal cord can dramatically alter the tuning of reflexes. These are called "upper motor neuron" lesions. Upper motor neurons differ from the motor neurons in the spinal cord, also called "lower motor neurons" that we have discussed. Upper motor neurons instead reside in the brain and they travel through tracts in the spinal cord to the anterior or ventral horn, where the lower motor neurons are found. These upper motor neurons and their axons thus stay strictly confined within the CNS. They are the central connections between the brain and spinal cord to impinge on lower motor neurons that in turn send axons into nerves. Recall that voluntary movements must first be activated from the CNS. Upper motor neuron problems, known as "lesions," damage descending connections and may arise from strokes, spinal cord injuries, MS or other causes, all CNS disorders.

Interestingly, upper motor neuron lesions cause deep tendon reflexes to be hyperactive. If the lesion is on one side, this may be a much brisker reflex than occurs on the normal opposite side. Why are interruptions of upper motor neuron pathways associated with overexcitable reflexes? The normal monosynaptic deep tendon reflex is dampened by descending upper motor neuron axons from the brain. When these axons are disrupted, the reflex is tuned up, or hyperexcitable.

What causes reflexes to disappear and is it important? This is now straightforward to understand. If a disorder or injury interrupts our monosynaptic three part connection, the reflex is eliminated. If the disorder damages the tendon sensory nerve, the spinal cord synapse, or the outgoing motor nerve, reflexes are interrupted. Thus, in our example of the quadriceps reflex, if a spinal disc compresses the third or fourth lumbar nerve root, the reflex on that side will disappear. The culprits here are the motor axons that control the quadriceps and exit the spinal canal at this level. By a disc compressing the exiting nerve root, both the sensory axons and the motor axons are disrupted and cannot relay the deep tendon reflex. A person with sensory polyneuropathy may have degeneration of sensory nerves widely throughout the body. While the motor axons may be normal, their abnormal sensory axons are no longer responsive to the reflex hammer. If the condition is severe enough, the deep tendons may be absent everywhere. Neurologists call this "arreflexia."

We mentioned other deep tendon reflexes that normal persons have beyond the quadriceps reflex, also called the "knee jerk." These include the ankle reflex that flexes the foot (foot bends down) when the Achilles tendon, the large cord at the back of the ankle, is tapped. Biceps, brachioradialis and triceps reflexes are all found in the arm. There is even a "jaw jerk," or more correctly a masseter reflex. The person holds his or her jaw apart and the hammer is gently tapped over the chin. The reflex causes the jaw to close. Fortunately the masseter muscle can be observed to contract a little but not enough to sever the fingers of anyone with their fingers in the mouth of the person being tested!

What happens if you are born without any deep tendon reflexes? This can occur, for example in someone with an inherited disorder of nerves. It is remarkable however that despite the arreflexia there may not be major disability. Sufficed to say that people lacking some or all of their deep tendon reflexes, without other neurological problems, are not usually troubled. More distressing are symptoms from the cause of the reflex loss. This might be a painful disc in the back or a sensory polyneuropathy that causes numbness and pain. Simply losing your monosynaptic deep tendon reflexes without any other deficits does not disturb your walking, and does not cause pain or weakness. Its something for neurologists to worry about.

Finally are there true, more complex, "polysynaptic" reflexes that do not require thinking? You might suppose that some politicians have these but in their case it may be the lack of thinking rather than the reflex that is the problem. Indeed, there are indeed a number of "classical" polysynaptic reflexes that "normal" people have. The abdominal reflex is a contraction of an abdominal muscle when a light pinprick scratches slightly over the surface on one side of the umbilicus. The anal "wink" is a contraction of the involuntary anal sphincter, when the skin around it is briefly scratched. This is not a favorite reflex for testing. Other new polysynaptic reflexes can appear in some persons with neurological problems. In persons with dementia for example, a small skin stroke over the base of the thumb causes the chin on the same side to contract, the palmomental reflex. Another polysynaptic involuntary reflex is the "grasp reflex," a tendency for newborn babies, or older persons with dementia to involuntarily grasp your hand. You might say to the person "Do not grab my hand" but instead, placing your fingers anywhere in their palm causes them to reflexively grasp. The grasp reflex occurs in adults following damage to the frontal lobe of the brain and in babies from incomplete frontal lobe maturation and myelination.

The overall message in this section is that reflexes are important but most are not the kind you might be thinking about. Another message is that reflex hammers are not designed to inflict punishment by neurologists but are tools that elegantly test specific parts of the nervous system. Deep tendon reflexes have a close connection to peripheral nerve anatomy and function. Just for fun, have a look at what kind of reflex hammer your neurologist uses. If its red and looks like a tomahawk, it's the American version. If it looks like a small tire or wheel at the end of a long stick, it's a "Queen's square" hammer named after the famous neurological center in London.

If it looks like a hammer with round rubber where it hits, it's a German type called a Tromner. Canadian neurologists are often unsatisfied with American tomahawks and prefer the English or German version. Many classically trained Canadian neurologists spent formative years in England, for example at the National Hospital at Queen's square.

## Nerves and muscles: The motor units

How many muscles can a nerve connect to? Each motor axon only connects to the fibers within one muscle. On a larger scale, only one nerve generally supplies a given muscle. A motor nerve branch would also not be expected to supply several muscles. This is helpful as you might not want one axon to have a branch connecting to a muscle that flexes a limb with another branch innervating an extensor. If this were the case, when the nerve was activated, both flexion and extension would occur simultaneously. This would make it challenging to walk! However, one motor axon can connect to several individual muscle fibers in a given muscle. The combination of one motor neuron, one motor axon, all of its branches, and all the muscle fibers it connects to is called the "motor unit" (Fig. 17). Motor units can be large or small, depending on the number of muscle fibers involved.

One motor axon can connect to a few individual muscle fibers or to hundreds. The extraocular muscles control eye movements. We considered them above under the discussion of MG. They include the superior rectus muscle that points the eye up, the inferior rectus that pulls it down, and lateral and medial rectus muscles that move muscles from side to side. There are two others called the superior and inferior oblique muscles respectively that coordinate with the rectus muscles (to respectively intort [rotate counterclockwise] the eye and move it down and in or extort [clockwise] the eye and move it up and out). All of these movements require very careful coordination between the eyes to avoid double vision. If the eyes do not move and align exactly, the brain cannot fuse the images into one. Extraocular muscles demonstrate precise control of movement. Their function is facilitated by the small numbers of muscle fibers connected to each motor axon. Compare this with a large muscle such as the quadriceps. We would not want our vision to depend on fine graded control of a large quadriceps muscle! Thus for the extraocular muscles around the eye, a single motor axon may only then connect to ten individual muscle fibers. In the quadriceps, as an example, one motor axon may connect to five hundred or more muscle fibers. In this instance, we want powerful, not finely graded movements.

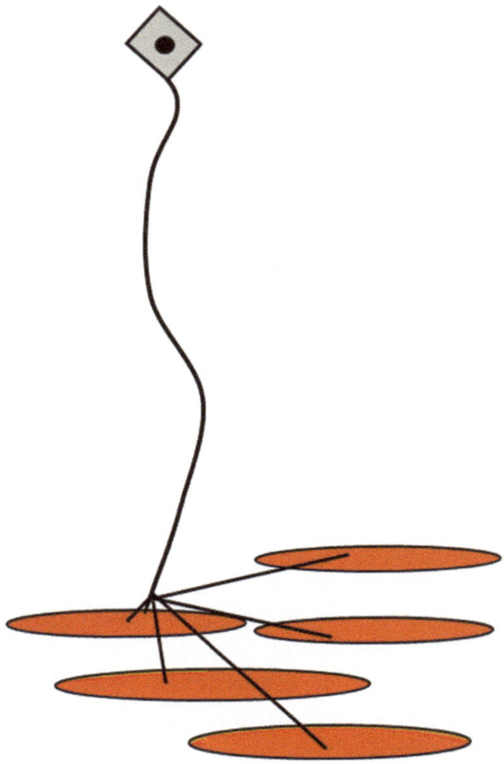

**Fig. 17** Diagram illustrating a single motor unit with a motor neuron, motor axon and branches innervating muscle fibers.

Motor units may enlarge in certain disorders. These can be recognized during EMG testing, described later. Why do they enlarge? If, for example, fifty percent of your motor neurons were to disappear over time, remaining healthy motor neurons send out branches to compensate. This is not difficult for motor axons to accomplish. In any given muscle, its motor units are intermingled. The target muscle fibers innervated by a single motor axon are not clustered, but instead dispersed throughout the muscle among those connected to other motor axons. The intermingled muscle fibers are connected to other motor units (Fig. 18). As a result of this structure, nearby motor axons have the capacity to send branches to neighboring muscle fibers that have lost their axon. In other words, in a disorder with incomplete loss of motor axons, remaining motor neurons and axons send out rescue branches. Over months to years surviving motor units have acquired greater responsibilities and enlarge, innervating their original muscle fibers

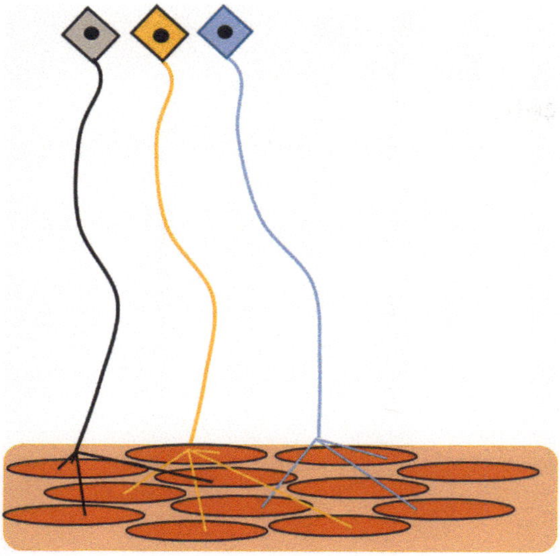

**Fig. 18** **Diagram illustrating several motor units innervating intermingled muscle fibers.**

and those rescued. A striking example of motor unit enlargement occurs in persons who have suffered from poliomyelitis in childhood, an infectious viral illness that directly targets and kills motor neurons. This is a paralytic disorder, also called "infantile paralysis." Survivors of poliomyelitis can make a remarkable recovery without new motor neurons to replace those lost. Muscle fibers are rescued by neighboring motor units that have survived polio, restoring lost connections between the spinal cord and muscles (Fig. 19). Of course outcomes are not perfect. When a long term survivor from poliomyelitis in childhood comes to the EMG laboratory to record motor units, the enlargement can be dramatic. Lesser degrees of motor unit enlargement can be detected in more subtle disease or during progressive disease of either motor neurons or axons. ALS, amyotrophic lateral sclerosis, another example, also exhibits enlarged motor units. Unfortunately however, this reprieve or rescue is only temporary since the disorder is progressive, leading to toward severe paralysis and death within 3–5 years.

Why do muscle fibers sprout and rescue their neighbors? This process, also discussed later, is called collateral sprouting. Despite the misconception that axon "wiring" is stable or unchanging, it seems that they await the opportunity to do something different. In this case, a muscle fiber without a nerve sends a "come hither" signal and it beckons new axon sprouts to connect.

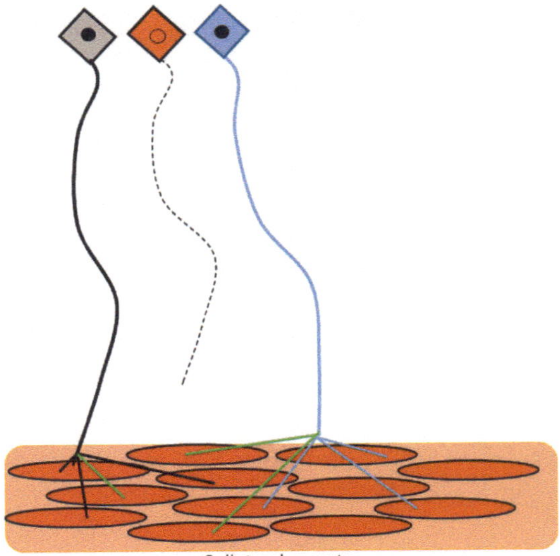


**Fig. 19** Diagram illustrating collateral sprouting of branches from intact motor axons (green) connecting to muscle fibers that have lost their original motor nerve innervation.

## Movement and Walking: putting your nerves out there

There is nothing more demanding of peripheral nerves than normal walking and perhaps running. Walking is a highly complex activity. It involves muscles, joints, and sensation, usually requiring a driver or brain. The interesting caveat to this is that animals can exhibit walking-like movement without a brain! The original physiological preparations that studied walking were called "decerebration" as discussed earlier, by separating the brain from spinal cord. Decerebrate experiments have showed us that rhythmic movements such as walking can be initiated without a signal arising from the brain.

Rhythmic walking-like movements arise from collections of neurons and their "wiring" in the spinal cord called "central pattern generators" or CPGs. CPGs form circuits within in the neurons of the spinal cord, that sets up a pattern, or a rhythm. How exactly it does this has been investigated by neuroscientists for some time. For example, a neuroscientist colleague, Dr. Patrick Whelan at the University of Calgary can ramp up a CPG by bathing a spinal cord in serotonin![21,22] So, if you are a little depressed, and taking medication that increases your serotonin levels, perhaps it helps your CPGs. It is an old maxim, without much evidence, that going out for a walk helps to lift depression, but I am not sure there is any "connection"! There is some controversy whether CPGs exist in higher mammals like humans. Decerebrate physiology experiments are not something human researchers would contemplate except perhaps in science fiction movies. So CPGs are fun to think about but what have they got to do with the PNS?

A CPG rhythm needs to instruct nerves to activate sequentially. In order to walk you must raise up your leg and knee, activating hip flexor muscles and their motor axons. Your other leg cannot engage in the same motion lest you fall down fairly hard. Your opposite legs must be straight as a board, beginning to extend in the opposite direction. This leg is called the stance leg, while the flexed leg is called the swing leg. Within this rhythm, you also have to pull your toes up so they do not scrape along the ground, at least in the swing leg. In the stance leg, you instead will push off on your toes. This is a vast simplification of the neuroscience of walking but it makes the point that motor nerves exhibit beautiful coordinated firing. In contrast, the sensory nerves might not seem to be important. However they also have a major role in walking. They tell you if you have overflexed or underflexed your muscles. They tell you if you have placed excessive strain on a joint or tendon. They also tell you whether you have walked into a puddle of water.

CPGs, walking and peripheral nerves are examples of the essential coordination between our nerves and signals from the brain and spinal cord. When nerves are widely disrupted there might be paralysis, preventing any kind of walking. Other abnormalities of walking, or gait depend on the type of damage they have experienced. For example, "foot drop" walking is easy to recognize. The person raises their swing leg far up above the floor to prevent drooping toes from catching the ground. Failure to clear the floor can cause a serious fall. If your nerves are partly damaged and cause leg pain, you may favor the leg or foot and spend less time standing or walking on it. This is called an "antalgic" gait. In an "ataxic" gait the legs are placed widely apart and turns are unsteady. This may arise from loss of myelinated axons that signal proprioception or where your legs are placed in space. Diseases of the CNS are associated with a host of other "classical" walking abnormalities, not discussed. What is remarkable is that among "normal" persons, how we walk can vary a great deal. Nerves and spinal cords have to achieve similar results whether they signal muscles in a 70 pound elderly woman or in a pro football player weighing upwards of 250 pounds with double the height and leg length.

## Exquisite sensation

The fidelity of our sensory nerves is astounding. The average weight of a mosquito is only 5 mg (!). Not only can we detect them landing but we can also tell if they are walking along our skin. Several unique qualities of sensation allow such sensitivity. The first is receptive field, or what territory of the skin activates a sensation. Second is how the sensation impacts the skin. There may be a brief "hit" on the skin as the mosquito lands and "hit" detectors are needed. These are called rapidly adapting sensory receptors. They respond to a change over the skin, a dynamic event. By deforming a small patch of skin, these receptors detect an event, or a "hit" and warn us that a bite might be next (if we realize it's a mosquito). How do we know if she (blood drawing mosquitos are female) is still there, preparing to bite? Our rapidly adapting receptors might not detect that. Once a transient dynamic impact is over, these receptors lose interest and do not generate a signal. For "static" sensations, for example, a band aid on your skin, receptors need to continue signaling that something is present. Thus, these receptors are called slowly adapting. You might want to know how large something is on your skin. For example, could this be a giant mosquito from Edmonton or Winnipeg? To signal this information requires sensory receptors with larger fields, perhaps less sensitive to fine touch on the finger, but sensing a greater expanse. As you might glean, receptors with small receptive fields are closer to the action, closer to the surface of the skin. Those with larger receptive fields are deeper within the skin.

To provide meaningful signals, our receptors must connect to nerve axons. This involves a hierarchy of responses. Fine nonmyelinated axons do not seem to require obvious sensory receptor structures. Their molecular makeup tells them when to signal. This is the case for pain fibers, also known as nociceptors, and for thermal fibers, sensitive to heat, warm or cold. To relay the high fidelity of touch, faster conduction properties are required. Recall that myelination adds properties to axons to allow much more rapid transmission than would be predicted for their size. Thus, touch receptors, vibration receptors and position receptors, also called proprioceptive receptors, are connected to myelinated axons. Some are connected to more than one axon.

How were these properties of skin receptors discovered? The physiological experiments that identified classes of receptor involved elegant and classical techniques. Recording electrodes were inserted next to sensory nerve trunks in a rat or other animal. This has also been done in humans! Fine mechanical stimulation with a rod could then be probed and the pattern of electrical response analyzed. This was called the firing pattern of the axons. Dynamic rapidly adapting touch fibers send off a wave of discharges that can be recorded and illustrated on an oscilloscope. Oscilloscopes are simply electronic devices that display rhythmic electrical patterns over time.

Before digital computers, mechanics could impress their customers by analyzing engine timing with an old fashioned analog oscilloscope! Newer oscilloscopes have a "digital" feel to them that is not quite as realistic as the older analog types!

Let us return our recordings. The "train" of discharges or action potentials is generated immediately as the stimulus, in this case the feet of a mosquito, distorts the skin. It reaches a maximum firing frequency when the rate of distortion is greatest, then slows and ceases once the skin is no longer moving. These are rapidly adapting signals and tell us that something has abruptly happened on our skin. Since different axons are connected to different receptors, this same "field" of skin will have other kinds of detectors lying in wait. The slowly adapting or "static" receptors may have a similar sized area of skin they are required to monitor. When our mosquito lands, it is not that acute bending of the skin or its movement that activates static receptors, but rather because the skin has now acquired a slightly different shape from the weight of this mosquito. Thus, a static signal detects that the normal equilibrium of the skin has changed because of the intruder. "Slowly adapting" static receptors send signals to fibers that increase in firing frequency less rapidly, but unlike dynamic receptors, they continue to fire signals as long as the skin is not normal. They do not adapt to the mosquito but they continue to signal it presence. In the physiology lab, rapidly adapting fibers are only sensitive briefly as the rod touches the skin. If the rod is left touching the skin, slowly adapting receptors signal that it is still there. Do we routinely sense every single touch stimulus on our body twenty four hours a day? The answer with slowly adapting receptors is probably yes, but its important to remember that the CNS is a large filtering station. If our brain were constantly bombarded by signals from every piece of clothing we wear, what we are sitting or standing on, or many other forms of data, we would be overwhelmed. What rises to our consciousness has to be of importance and worth thinking about. Our brains are highly selective as to what part of this information overload is relevant and important to focus on. If we are inattentive or drowsy, we might ignore most of it!

I was recently at a meeting that inadvertently modeled how sensation and brain filtering may work. The meeting room was set up so that all of the participants had a microphone on the desk in front of them. Approximately fifty participants were present. However, the microphones mentioned were quite sophisticated and emitted a small red light in response to a voice or even a random sound close by. The red light indicated the microphone was now on. This was very convenient because it meant that anyone wishing to ask a question or make a comment had simply to talk into the microphone. When your voice was detected, the small red light came on and you could be heard throughout the room. When quiet, the red light and microphone were off and background noise was kept to a minimum. Considered as a whole, the meeting room acquired an interesting dynamic in that the numbers of red lights depended on how controversial a topic under discussion was. During a boring presentation, there were few questions and few red lights. However, some did come on apparently randomly, briefly and sporadically. These were triggered by someone shuffling their papers too close to the microphone, coughing or drinking water. When something more interesting was presented, several dispersed or clustered red lights appeared and stayed on as the participants completed their question. Others had something to say to their close neighbors and red lights appeared for sidebar conversations. In a similar way, random activation of sensory receptors throughout the body likely occur throughout the day and night. However, your brain is not really interested in these events and they only rarely pop into consciousness. However, when something that matters happens, activation is more persistent, may involve neighbors and may light up other parts of the brain not unlike my meeting room. The signals warrant your close attention! Sensory receptors in the body take this to yet another level. When something is important, not only does the receptor stay active, but the frequency of its discharges also ramps up. This represents a more sophisticated response than a receptor light flashing on and off. During its filtering role, the brain interprets where the signal comes from, how many signals are struggling for attention at once, how persistent the signal is and how frequently the AP discharges are! Going back to our meeting room, we might design a system of flashing red lights so that the importance of the question might make our light blink on and off quickly! I am not sure I would enjoy that kind of meeting. Everyone thinks their own question is of utmost importance.

How a given sensory receptor, continuing with our main examples from the skin, is activated by fine touch is remarkable. This is the time to introduce some additional biology over how this occurs (Fig. 20). Touch is not straightforward! By touch you might detect the texture of an object. Is it a wool coat or a rubber raincoat? Your nerves can tell the difference. Is it something moving or something quite still on your skin? Is it pleasant or unpleasant?

Touch receptors have a complicated name. They are called low threshold mechanoreceptors, or LTMRs for short.[23] They are low threshold because

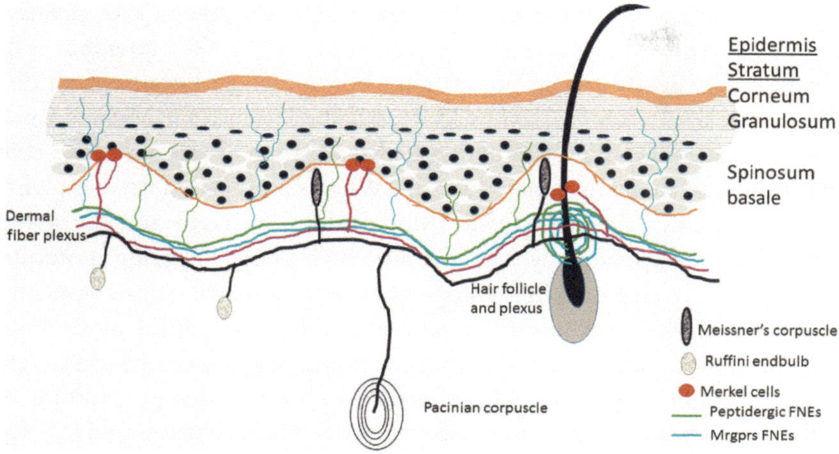


**Fig. 20** Diagram illustrating sensory axons and sensory organs in normal hairy skin. Sensory axons include nonpeptidergic Mrgprs free nerve endings (blue) extending further in the epidermis and peptidergic free nerve endings ending deeper in the epidermis (green). Merkel discs are located at the epidermal-dermal junction whereas Meissner's corpuscles, hair shaft innervation, Ruffinian endbulbs and Pacinian corpuscles are found in the dermis.

we are talking about touch, not being hit by a sledgehammer! Low threshold simply means that not much is required to set it off. Remember our mosquito! LTMRs cover a variety of types of touch. For example, a large part of the skin may have been touched or a small area. This is called the receptive field. The touch might be very transient or off and on, so a receptor has to be able to turn off and on rapidly, i.e., rapidly or slowly "adapting," as described earlier. The touch could involve moving hairs or be directed to nonhairy skin, called "glabrous." Overall as many as 17000 mechanoreceptors may be present on the human body![23]

Let's examine these touch receptors further. In the base of the epidermis of the skin, which is very close to the surface, are Merkel's touch domes or discs that in turn contain Merkel cells. Friedrich Sigmund Merkel (1845–1919) was a German anatomist who first depicted red images for arteries and blue images for veins in addition to the interesting touch receptors named after him (https://en.wikipedia.org/wiki/Friedrich_Sigmund_Merkel)! Together these discs and touch domes are LTMRs that are slowly adapting with a small receptive field. Their small receptive field makes sense since they are positioned superficially in the skin. Merkel's discs are comprised of clusters of Merkel cells called Merkel domes. One axon can connect up to seven domes in which there might be up to 150 cells apiece. This capacity allows one axon the capable of responding to over 1000 individual cells. However, the diversity of responses does not end there because each disc can be associated with two axons. "Haarscheibe" domes are Merkel touch domes in hairy skin, about 0.2 mm wide that are also clusters of Merkel cells innervated by a single myelinated axon.[12] What is clear is that there is plenty of overlap among axons and cells. Merkel discs are rigid and the axons within them are activated by static pressure such as our lingering mosquito. Indenting a tiny area of skin is sufficient to activate them. Some have suggested that Merkel cells might operate by sending a chemical signal to its axons, resembling an unusual synapse, but found within skin. Merkel's cells are also thought to be uniquely sensitive to corners, edges, curvatures or anything that indents the skin to a depth of 15 microns or less and 0.5 mm apart. That is not very much! Erstwhile they mind their own business and do not pay attention to events that are not in their immediate zone. They are also not interested in stretch sensations. Needless to say perhaps, greater numbers of Merkel cells are found in sensitive parts of the skin like the fingertips and lips. Merkel cells may also be associated with the base of some hair follicles. Unfortunate mice who happen to be genetically engineered to lack Merkel cells cannot tell texture with their feet.

Next on the list are Ruffini endbulbs that are found deep in the dermis and measure 1 mm long and 30 microns wide.[12] As you might guess from their location, these LTMRs have wider receptive fields. In fact their receptive fields are thought to be five times larger than those of a Merkel disc. Ruffini endbulbs are also slowly adapting, generating APs as long as the skin is deformed but they seem most sensitive to stretch or folding of the skin. They also respond to simple indentation but are not as sensitive as the Merkel disc. Ruffini endbulbs have a unique structure with layers of cells surrounding a fluid center that houses single myelinated axon poised to signal. Multiple small terminal branches emerge from the parent axon to mingle with the resident capsular cells. Ruffini endbulbs also inform the brain about the posture and positioning of the hand, not by relying on joint movements but by detecting the degree of skin stretch. Our ability to determine where our hands are by sensing skin stretch, joint position and tension of tendons and muscles is extraordinary! Redundant systems like these might be important forms of sensory backup or reinforcement. Of course, some people deliberately have their facial skin tightened to look younger. How the Ruffini endbulbs feel about that intervention is unknown! Less well characterized variants are called endbulbs (endbulbs of

Krause or Golgi-Mazzoni) in which axons and their branches wind about within capsules. These are found in mucous membranes or within the dermis of glabrous skin adjacent to orifices.[12]

Rapidly adapting dynamic skin receptors detect skin movement, tapping or vibration. Small receptive field rapidly adapting receptors are called Meissner's corpuscles. They are located immediately beneath the epidermis in the outer dermis and measure 80 by 30 microns, oriented perpendicular to the skin surface.[12] Meissner's corpuscles are prominent in the finger tips and soles of the feet. They are filled with fluid and contain epithelial cells or lamellae, that sandwich up to three myelinated axon terminals. The axons intricately weave among the lamellar cells, making them especially prone to slight distortions.[16] It is thought that the proteins within the corpuscle deform rapidly to sense the stimulus, then rapidly recover, or adapt. Each protein movement generates an "on" or "off" signal. A transient stimulus (our mosquito leg briefly alighting!) directly over this receptor sends a ripple through the skin and corpuscle that is magnified by this structural arrangement and activates the axon within it. They can detect vibrations, but only around 40 Hz, outdone by the Pacinian corpuscle described next. Since each parent neuron supplies several Meissner's corpuscles, signals are summated from several receptors and it is also estimated that a given axon can supply up to 80 corpuscles. As additional receptors are stimulated, the signal directed to the brain is amplified. In addition, Meissner's corpuscles are sensitive to movement across the skin. They may also be important for maintaining a tight grip. For example, when your grip starts to slip, these sensitive skin receptors first detect the movement and allow an immediate adjustment. For this problem, slowly adapting receptors would be useless. You might have dropped that dish you were holding by the time these receptors noticed that the skin indentation had changed! There are no receptors that relay embarrassment. This comes from the brain. Finally, parallel processing from the Merkel cell and Meissner's corpuscles allow humans to read braille![24] Bolton, Winkelmann and Dyck counted Meissner's corpuscles in humans.[25] On the tip of the finger, younger persons (ages 11–30) had a mean of 24 Meissner's corpuscles per square millimeter. After age 50, the density declined by over 50%. The bottom of the large toe, on the other hand had densities that started at 11.0/sq mm and also declined by a similar amount over the age of 50. Unfortunately our exquisitely wired sensory system, like other parts of the body, gradually declines in sensitivity with age!

Finally, yet deeper in the subcutaneous tissues are the Neurologist's favorite receptors, the Pacinian corpuscles that detect vibrations, but over a wider receptive field. One estimate is that the human hand has 2500 of them. These corpuscles are also more sensitive than Meissner's in that they detect finer motions. Pacinian corpuscles look like microscopic onions measuring 1.0 by 0.7 mm within the deeper dermis of the skin.[12] The ultrastructure, or fine anatomy identified by an electron microscope (EM) of the Pacinian corpuscle was beautifully described by Peter Spencer and Herbert Schaumburg from New York,[26] also well known for their seminal work on neuropathy following toxin exposure. A single myelinated axon enters the corpuscle and after one further myelin segment, sheds its myelin. The nerve within the Pacinian corpuscle is surrounded by stacks of thin hemicylindrical cells that give rise to long thin crescent like profiles called hemilamellae. These are what form the "onion skins." The end of the nerve, or "ultraterminal axon" in the end of the corpuscle called is enlarged and forms fine fingers that connect to the hemilamellae. Interestingly, the enlarged axon endings resemble growth cones of regenerating axons that we will discuss later. To quote Spencer and Schaumburg "The structure and location of the axon processes appear to be eminently suitable for detecting pressure transients transmitted through the outer core of the corpuscle". Alone, these delicate and sensitive receptors are specialized to their task and could not be relied upon to provide an overall appreciation of what object might be in your hand. They are sensitive to a variety of vibration frequencies from 60 cycles per second (Hz) to 400 Hz. While tuning forks used by most Neurologists are generally rated at 128 Hz, some have argued that 256 Hz should be used since the most sensitive frequency detected by Pacinian corpuscles is 250 Hz! At that frequency, a vibrating movement of 1 micron can be detected. Irrespective of whether you adhere to the 128 or 256 club, the importance of the tuning fork is not just the fact that a select class of receptors is sensitive. A test of vibration sensation in the toe for example, tests the integrity of a great deal of the nervous system: the Pacinian corpuscles in the toe, the myelinated nerves that supply them, the neurons in DRG that support the myelinated axons and all of the connections from these nerves to the cortex of the brain. At the cortex, the last stop of the pathway, we are made aware of the sensation of vibration. A disorder that interrupts any segment of the pathway results in an inability to feel the neurological tuning fork.

An interesting new finding concerns the types of axons that transmit touch. LTMRs are classically thought to be associated with myelinated axons. That touch requires myelinated axons has been classical neurological teaching but in fact, it is not so. Unmyelinated LTMR axons also exist and are numerous in hairy skin and are closely associated with hair follicles. Unmyelinated or C fiber LTMRs may transmit sensations that are called emotional touch.[27] These are pleasant sensations, caresses or even perhaps tickling! Here is proof that not all emotions strictly arise from the brain!

We have discussed the receptive field, or area of skin connected to a given sensory axon and how it is larger if the receptor is positioned deeper in the skin. It is also larger if the receptive field is in a less sensitive area, for example the forearm compared to the finger tip. The receptor density, has great importance in the perception of sensation, i.e., how many sensory receptors, and axons that connect to them, serve a given patch of skin. Its patterning over the body accounts for the variation of skin sensitivity we experience. For example, density is highest on the lips, genitals and finger tips for both Merkel's and Meissner receptors. Having a higher number of receptors in a given patch of skin helps to specialize parts of the skin for different functions. For example, it would not make sense to have large numbers of sensory receptors on the back skin of humans because we do not explore our environment through our back skin! Alternatively, it is essential to have high densities where we interact with the environment and manipulate it using our hands and fingers.

How sensitive is the skin? Measuring fine skin discrimination to touch is a useful neurological technique. The measurement of two point discrimination uses an adjustable probe with two pointed but blunted tips. It measures of how far apart two objects on your skin have to be for you to recognize that there are two objects and not one. The test relies on receptor density. With eyes closed, the subject is asked to tell whether they sense one or two points on their finger tip. At random intervals, the examiner uses one point, or two points spaced at different distances apart. When two points are less than 2 mm apart and applied on the fingertip, it is generally felt as one stimulus by normal healthy persons. As the distance widens between the points to 3 mm subjects will feel two points. If there is a loss of touch sensitive axons in the skin this distance expands, indicating decreased two point discrimination. In normal persons, two point discrimination in less sensitive areas of the skin is wider apart, such as over the back where it measures 30–40 mm. Thus, if two fine objects touch the back, and they are closer than 30 mm apart, you only feel one touch. Try it! What determines the spacing of skin receptors is worth further investigation but it is largely unknown. For example, it may be that axons send signals to their neighbors to grow elsewhere. Axons may be territorial, actively sending destruct signals to branches from other axons. An older idea is that the skin provides molecules that offer support for its axons, a property known as trophism from locally available growth factor proteins. If only a limited supply is available, fewer axons remain to fulfill their roles as sensory signal beacons.

The most recent chapter in the touch sensation story is the discovery of "Piezo" channel proteins that detect movement. These convert a mechanical movement into an electrical charge that can be transmitted up the nerve. Touch and movement depend on a local force causing a temporary deformity of a protein known as Piezo2. The proteins are configured so that their three dimensional structure shifts whenever the skin moves. Upon stretch, these channels have the fascinating property of opening an ion channel within them. Imagine Moses parting the Red Sea long enough for he and his people to pass through. The idea is that the protein is delicately folded where it sits in the membrane. It is then twisted and changed by mechanical pressure: direct touch or pressure, suction on the membrane or shear forces. The skin bends, the cell membrane bends, the protein bends, and behold a passage is opened for ions! Two consecutive papers in "Nature" have recently described the relationships between the structure of the Piezo2 "tactile channel," its Piezo1 relative and their function.[28,29] It's a highly complicated but very elegant arrangement revealed by a high resolution imaging technique called cryo-electron microscopy and tested by another technique called atomic force microscopy. The protein is shaped like a propeller with three arms that are evenly spaced around a curved bowl or dome and a cap. The idea is that a force flattens the dome and opens its ion channel to cations (positive ions). The propellers may act as springs to quickly bounce the structure back into its dome shape later. The electrical charge caused by stretch then activates or depolarizes a sensory axon terminal to generate an AP transmitted to the CNS. Here is a unique way that a force or touch on the skin can be coded into an electrical charge.

Piezo channels are a part of a larger group of proteins known as mechanically activated channels. One member of the family, Piezo2, is found both in the Merkel cell but also on the sensory nerve connected to it. Genetically engineered mice lacking Piezo2 in Merkel cells, do not respond to touch properly. Why both axon and the Merkel cell seem to require the channel is not yet understood but intriguing. We are learning that axons, neurons, SCs and sensory receptor cells are not islands unto themselves. They have critical ways to transfer molecules between each other. Whether this includes a Piezo protein is unknown. An important part of this story is that families with genetic abnormalities of their Piezo proteins have neuropathies, like the mice, and also have impaired touch sensation. For example, two family members were recently described in the New England Journal of Medicine[30] who had genetic loss of functioning Piezo2 channels. Careful testing of the sensation of these individuals revealed surprising abnormalities. There was loss of vibration sensation, loss of touch sensation on nonhairy skin and loss of position sensitivity of their joints. They had difficulty walking, problems reaching and an inability to maintain their balance with their eyes closed. Sensations to pressure, pain, temperature and itch were preserved. In addition they were normally sensitive to hair stroking. Although both subjects had difficulties with movement and posture, they compensated remarkably well using their other forms of sensory input. Both patients also had delayed motor development including fine motor skills. Finally these patients also had early problems in head control, breathing and feeding and developed skeletal abnormalities such as hip dislocation and scoliosis (spinal curvature). If Piezo channels are only found in sensory axons and Merkel cells, how is it possible to have all of these other difficulties? While the answer is unknown, we are aware that normal sensory wiring during development is highly important in supporting motor function development. In Chapter 6 we will learn about a patient without any sensory neurons who could not walk! These are very striking examples of how sensation has a major impact on other parts of the body.

The next layer of sensation is offered by hair! Human hair provides very little heat conservation compared to our primate cousins. We do not fare well without added heat protection, particularly in Edmonton where I write this. However, hairs are unique sensory detectors. In rodents highly sensitive and sophisticated whiskers known as vibrissae have evolved beyond what humans possess. Humans do not have them despite some that sport large buckethandle moustaches. So interesting is the sensory coding by specialized vibrissae hairs that physiologists have spent careers investigating them. Each hair might be encoded with sets of neurons patterned in a "barrel-like" pattern in the brain that detect their actions. Barrels are important sites of brain sensory processing, used by other modalities like vision. However let us return to nerves and the PNS.

In humans, the role of our hairs in sensation is less impressive and we have rudimentary whiskers. In fact many people shave them off of their face or legs! In humans there are three separate types of hairs, each having sensory receptors in their hair follicles. Tylotrich or guard hairs are large hairs as may be found on the scalp. Their follicles are associated with rapidly adapting receptors. Thus a transient bending of the hair, from wind blowing through it, evokes a discharge and a sensation. Think of hair shampoo commercials with locks rustling as the models shake their head. However, if your hair is matted down from wearing a hat in the heat, so called "hat head," there might not be a sensation from these hairs. You may not like how it looks but it does not sense movement the same way. Other receptors in the skin might detect the pressure from hairs forced down under the hat. Guard or tylotrich hairs are coarse, found over the body, albeit largely on the scalp in humans. Down hairs that are smaller and softer, make up fur in animals and they are similarly endowed with rapidly adapting receptors. They are sensitive to stroking. At least some of these receptors have the Piezo channels we described above. Many nerve endings in hair follicles are called "lanceolate" because of their distinct architecture that resemble leaves. They also have the appearance of a lance or spike around the base of the hair follicle. As the hair bends, tension between the hair follicle and lanceolate endings causes the rapid onset of discharges. One estimate is that these receptors can detect one mm of movement.[16] Eyebrow hairs in humans are thought to be particularly sensitive! Part of the lanceolate "leaf" tip, differs from other parts, setting up unique signals that travel up along the nerve. What may be evident is that any given hair follicle may activate different sensory nerves depending upon the type of stimulation. A lingering gentle touch to the skin might stimulate slowly adapting axons, whereas a quicker poke stimulus activates rapidly adapting axons including those of the hair follicles. A breeze might only activate axons from a series of follicles with hairs that bend together. Merkel cells are not only part of the hair follicle but they are also associated with the skin adjacent to hairs so that their movement can be detected by changes skin pressure. Recall that Merkel cells are slowly adapting, endowing hairs with yet another form of sensation. Some have suggested that the combination of rapidly adapting and slowly adapting hair sensations also allows a form of parallel processing input from hairs eventually to the brain. These sensations are all encoded by a "piloneural" complex, with axons forming a collar around the base of the hair shaft just deep to the sebaceous glands of the skin.[16] A sensory axon might only connect and respond to one hair follicle. Or, one sensory axon might also connect to a series of hair follicles. Occasionally a single hair movement might not interest the brain at all. However, some axons connect to a series of hairs, all of which need to be moving to activate an AP. In this way, a stroke over hairy skin is recognized by the sensory fibers long before the signal is decoded by the brain. Some of the axons connected to hair follicles are medium in size and myelinated, called $A\beta$, others smaller and are called $A\delta$. These latter, smaller axons connect to finer hairs on the body.

Awareness of the position of a body part or its movement in space also relies on sensory receptors. These are called "proprioceptors." They are attached to the thickest and most heavily myelinated of the sensory axons and have the fastest conduction velocities. Their axons are called Aα. This differs from the touch receptors Aβ mentioned above that are one step lower in size and myelination. Receptors sensitive to limb movements can also be called "kinesthesia" receptors.

Small specialized organs in muscles known as muscle spindles, acting as stretch receptors, are important proprioceptors. These are found on separate muscle fibers that run in parallel with normal fibers. Spindles encapsulate and separate groups of 2–14 muscle fibers that contract or stretch in tandem with the muscle they are embedded within.[12] To simplify and summarize, most of our muscles have two parallel systems of muscle fibers: "Normal" fibers contract to move the limbs, eyes or jaw. Within these are the smaller groups of encapsulated specialized fibers, the muscle spindles. Spindle muscle fibers have stretch receptors and are also called "intrafusal" whereas the larger number of normal muscle fibers outside spindles are called "extrafusal." Intrafusal muscle fibers also have their own motor neuron supply from specialized "gamma" motor neurons and axons. By setting the degree of tension of intrafusal muscles, the gamma motor neuron AP firing pattern controls how much the muscle spindle gets stressed when all of the muscle fibers are being stretched. For example, if the biceps muscle is stretched, but gamma motor axons keep the intrafusal spindle muscle fibers contracted, the impact of the stretch is more intense. Try to have someone pull your arm out while you are actively trying to flex your biceps muscle. This can be painful! Gamma motor axons thereby tune the sensitivity of the spindles to overall stretch. To sense this stretch, intrafusal spindle muscle fibers possess a dense supply of nerve fibers around them. The "wiring" of myelinated axon terminals coiled around intrafusal muscle fibers is actually beautiful to behold and the story is yet more complex. Intrafusal muscles have colorful names like "nuclear bag" and "nuclear chain," depending on how their nuclei are clustered or lined up and how their nerve supply behaves. Beyond the novel name and appearance, bag and chain intrafusal fibers have different physiological properties, not unlike the skin. Some types of bag fibers (bag2) are dynamic and respond to movement whereas other bag fibers (bag1) and chain fibers respond to static positions of the muscles. There are primary large myelinated sensory axons that supply the midportion of the intrafusal muscle fibers and wind around them, called annulospiral endings. Secondary endings are from slightly smaller myelinated sensory axons and they are attached near the ends of the intrafusal fibers with "flower spray" endings.[12] It is wrong to assume that muscles are simple bundles of contracting cells! They have significant complexity closely linked to that of their motor and sensory axon supply.

When a tendon is stretched, specialized receptors signal yet another way for the body to recognize a change in position. These are tendon afferent sensory fibers that we discussed when covering deep tendon reflexes, also called Golgi tendon organs. They have a capsule, with collagen (a connective tissue protein that forms tendons) fibers that is innervated by one or more myelinated axons forming terminal branches with swollen "grape-like" endings known as varicosities.[12] Like muscle spindle axons, they detect stretch, stimulated by tension of the tendon.

A third position sensor comes from the actual joint capsule that is deformed as we move or position ourselves. Joints have Ruffini, Pacinian, Golgi tendon organs and free nerve endings innervating them.[12] Of course, all of the skin receptors we have already described above are critical to joint movement as well. These include skin "field" receptors, yet another name for responsiveness to stretch by a single receptor or an ensemble innervated by the same axon. Stretch occurs over the joints when we move them. As already alluded to, skin axons also detect what we are walking on or in. The remarkable fact is that all of the signals from a variety of sensors and nerves send a barrage of information that is filtered and interpreted by the brain. As a result, we can detect when we walk on a sandy or muddy bottom of a lake beach. One can imagine how it would take time to adapt to low gravity, such as that found on the moon. All of these signal inputs in this scenario would provide new information and the brain is required to adapt them to a new environment! Similarly, neuropathies often damage the long endings of the nervous system first. Loss of balance is one of the first symptoms experienced by neuropathy patients. There is loss of sensitivity to the floor and if the eyes are closed, these patients are unable to predict where their legs are!

Sensitivity to temperature and pain are closely related. Both types of sensation are carried by small "bare" nerve endings of nonmyelinated fibers. These sensory terminals do not appear to have unique structures attached to them, like Merkel's discs or Pacinian corpuscles (but they may associate with unique Schwann cells—see the note added in proof in Chapter 1). Earlier, we discussed whether we are capable of recognizing the speed of conduction transmitted by our axons. With one exception, this is unlikely. An exception may involve the types of pain sensation we experience. Some signaling uses myelinated axons thought to convey acute pain whereas other pain signals rely on unmyelinated axon conduction. Using the idea of "slow" versus "fast" pain transmission, a British neurologist, Henry Head (1861–1940) and his colleagues established descriptive terms for pain. They described "epicritic" pain as immediate, fast, carried by smaller myelinated axons that conduct around 20–30 ms/s. This is the acute or immediate pain that triggers you to pull your hand away from a fire. In contrast, "protopathic pain," thought to be transmitted by unmyelinated axons, was described as slow, delayed and burning. This is the pain that builds up when you have actually burned yourself. We now know that this description was an oversimplification but it does help one remember different types of pain sensation.

The sensations for heat and cold are both painful. In Edmonton we argue over which of the two is actually more painful! Temperature sensing nerves are structured to signal this important warning. Some axons respond to more moderate cooling or to warming, sensations that are not painful. This would be like comparing the cold during February in South Carolina to that of Edmonton. Overall, thermal sensations can be signaled through small myelinated axons while others use nonmyelinated axons. Earlier we discussed small unmyelinated "C" afferents of the epidermis including peptidergic free nerve endings and nonpeptidergic mrgprs axon free nerve endings that travel slightly further out into the epidermis. Both types of axon signal and transmit pain. Some pain receptors, called nociceptors, share sensitivity to more than one stimulus. Thus thermomechanical sensory receptors respond to both heat and mechanical pain. What determines an individual axon response depends on its protein content including a fascinating group of finely tuned molecules. For example, mechanical nociceptors that strictly respond to mechanical injury are wired to small myelinated axons. The myelinated wires are called Aδ fibers, smaller than Aα and β axons that connect vibration and touch receptors described above. Yet others are called "polymodal," because they respond to a variety of stimuli and are wired to nonmyelinated C fibers. These terms seem to have been created for no more sophisticated reason than their order in the alphabet. In any event, the smaller fibers we have described connect to smaller neuron perikarya and large fibers to large neuron perikarya. Why should there be two separate wiring diagrams for pain? Pain is critical for evolutionary survival. Why we need to have "fast" and "slow" pain is also an important philosophical question. It may be that one requires a rapid motor response and the second reminds the animal that all is not well, a serious injury has occurred or is in progress.

There are nociceptors in the skin and body that do not signal without a major push, the "silent" nociceptors. The idea is that these free nerve endings, embedded in skin or other tissues, await the time they may be called to action. Some might only respond to severe mechanical stimuli like a crush injury. Others await a chemical signal. Several chemical signals can activate sensory nerves; some causing itch and others causing pain. The hydrogen ion activates sensory channels known as "acid sensing channels" (ASCs). Subtypes of ASCs exist but they have a diversity of protein structure and do not sense different types of acid. How one perceives acetic acid (vinegar) or muriatic acid (hydrochloric acid) involves the same channels but their content of hydrogen ions determines how strongly they are sensed. For example, while you might want to ingest acetic acid, muriatic ingestion would not be a good idea. By the time we have this discussion, we are far beyond ASC sensors or nerves and into severe and raw tissue damage caused by powerful acids! The smaller hydrogen ion concentrations discussed here activate only a few nociceptors and do not cause this kind of damage!

In recent years, understanding how temperature is coded by nerves has progressed. We now know that a series of temperature sensitive proteins are present in skin and other axon terminals. While under the microscope only "bare" nerve endings characterize these axons, their content of "TRP" proteins determines their role. TRPs are "transient receptor potential" ion channels on axon membranes that allow $Ca^{2+}$ or $Na^{+}$ to enter and APs to be generated. At last count 28 distinct TRPs have been discovered, but only a few key ones need to be described here to explain how they work. The classical and first described TRP channel was TrpV1, also called the vanilloid receptor. This receptor is exquisitely sensitive to the hot spice capsaicin, the active ingredient of chili peppers. If you suddenly bite into a hot chili pepper, you will recognize the kind of stimulation activation this channel brings about! After exposure to capsaicin, TrpV1 channels allow calcium to enter into the axon, in turn generating a series of APs. The intensity of the sensation you notice depends on how many axons were activated and for how long. This of course depends on how many hot peppers were in your mouth, exactly how "hot" (how much capsaicin they contain, not their temperature!) they were and for how long you kept them there! Fortunately TrpV1 channels can become desensitized and experienced diners are no longer troubled by this kind of exposure. Some foodies may actually seek out these kinds of foods and spices. "Death sauce" or Louisiana Hot sauce comes to mind. TrpV1 channels are also sensitive to heat from 35 up to 50°C. This is why these foods seem "hot" even though they might be at room temperature. With this in mind, it is quite accurate to speak of "hot" foods given the fact that the same protein channels and nerve fibers are both activated by capsaicin and heat. On the other end of the TRP spectrum are the TRPM8 channels, sensitive to cooling down to 10–20°C and to "cooling" stimuli such as acetone, menthol and eucalyptol. The "M" in TRPM8 stands for menthol! There are TRPA1 receptors that detect yet cooler temperatures and are activated by wasabi and garlic, and TRPV2 receptors that can detect painful heat up to 60°C. Other members respond to intermediate temperatures.

In neuropathies, diseased sensory axons may change what signals they send. Some patients describe that heat feels cold or cold feels hot. Thus, when these persons dip their feet into water, they may experience the opposite sensation to what is expected. If the water is hot, it might feel cold to someone with neuropathy. If the water is cold, it might feel hot! There is a term for this, "paradoxical temperature sensitivity." How this develops is still debatable. One explanation may be that one type of fiber, for example cold sensitive axons with TRPM8 proteins, degenerates leaving only TrpV1 axons intact. Thus the brain interprets any activation whatsoever as a heat sensation. Another interesting possibility is that any given axon shifts its protein content. Axons that formerly had mainly TRPM8 proteins, change their properties and now produce TrpV1 proteins, sensitive to heat. Inappropriate feelings of cold when the stimulus is warm could arise from the reverse of this-TrpV1 proteins get replaced by TRPM8 proteins. The fact is that we also do not know many warm sensitive fibers or cold sensitive fibers should normally be present on the skin. Careful measures of how sensitive people are to either warm, cool or painful heat or painful cold tell us there is a "normal" range that almost all people fall into. Finally despite the fact that this book is about "wiring" we do have to admit that the brain determines the final interpretation. So, if you are in Edmonton, what is cool might be painfully cold if you are a native of Florida and not a Canadian. This does not mean you are a more robust human being, just that you have a different thermostat built deep inside your brain. The hypothalamus, a structure in the center of the brain below the thalamus (hence "hypo") that controls body temperature, may be the culprit. Some Edmontonians feel they should carefully reprogram their brains by moving to Florida!

## Considering taste

Having good taste is viewed as a sign of refinement. Today, this might mean choice in clothes, automobiles, art work, friends, books or other things. Tastes are preferences. However, taste actually involves the peripheral nervous system. A "foodie" is someone who considers themself to have good taste in choosing meals. Significant biology is behind all of this. We humans have expanded what is a wiring and transmission problem to a range of preferences that most brains have difficulty fully grasping.

The sensations that trigger all of this human thinking arise from exquisite wiring and signaling. Taste is not among the "ordinary" senses but is called a "special sense." While some of its wiring looks familiar, its receptors are not. The new biology of taste buds that house these receptors has been well summarized in a "Nature Reviews in Neuroscience" paper by authors Roper and Chaudhari.[31] Its important to distinguish taste "buds" from taste receptors. Taste buds are a collection of specialized organs, containing multiple cells, found on the tongue. Interestingly they are also found on the inside of the mouth. In older times, when hapless victims had their tongues removed for their transgressions, they might still be able to taste. Of course, the rest of the meal experience might not be very attractive without a tongue.

Taste buds have tips that project outward to interact with food or fluid. Within these tips are cells that contain the taste "receptors." These receptors, in turn, are specialized endings connected to nerves, not unlike the Merkel discs and Meissner's corpuscles of the skin. Taste receptors use specific molecules known as "G" proteins. While G proteins have no relationship to "G-men" of the FBI, they are found throughout the body in various forms and are activated by many molecules and medications. Their proper name is G protein-coupled receptors (GPCRs), proteins with subunits that span from the surface of the cell to inside the cell. When a specific molecule binds to its outer subunit, the protein is activated through a change in its shape and its inner subunit signals the event to the inside of the cell. In more detail, since the protein spans across the membrane, this binding and conformational change of the protein now allows it to interact with neighboring intracellular proteins waiting for action. In taste receptor cells activated GPCRs set up a biochemical signal, transmitted through the cell that eventually activates sensory endings.

Taste buds are complex structures that look like a garlic clove with layers of over 50 elongated cells that point to a tip. Within them, are specialized cells called Types I, II, and III. Any given taste bud might have several of these cells all mixed together. Some cell types are more common than others. Type I cells comprise about half the number and provide support to other cells of the taste bud. They do not actually transmit taste, but support the cells that do, not unlike SCs. Support is everything! Type II cells comprise about a third of the taste bud cells. They actually do transmit taste but the type of taste depends on their exact type of GPCR. Remember that whole families of related but differently behaving GPCRs exist, an extended family! Most Type II cells are specialized to contain only 1–2 individual types of GPCR for taste. However the taste bud, by housing many types of cells, is sensitive to many tastes! Type III cells represent only 2–20% of the cells and their numbers in a given taste bud vary. Here is an exception since these cells do not have GPCRs but instead use synapses! Most taste cells using GPCRs have closely apposed axons to transmit taste signals but Type III cells appear to synapse with their axons. There are also other connections within the taste bud that influence how their signals are produced. Some use molecules such as ATP, serotonin or GABA as neurotransmitters within the taste bud. Thus you can appreciate how all of this complexity contributes to the sensation of taste: multiple cells, multiple GPCRs, synapses and neurotransmitters all together in a single taste bud! In some ways they are like miniature "foody brains."

Just to be clear, not all of the nerves of the tongue are used for taste. Like skin, the tongue also has other receptors, one of the reasons that it is so sensitive. If you truly burn your tongue with a hot (temperature) food or a beveridge, you will experience pain rather than a stimulus that activates taste buds. If the burn is severe enough you actually might destroy taste buds. The tongue thus has pain receptors but also touch receptors that are activated by other stimuli. Of course, when you are eating, your tongue touches the food and can tell its temperature, shape, texture. You might have a habit of being annoying by rolling food around in your mouth. This happens because you can detect food details, using LTMRs, the receptors discussed in the skin with sensitivity to either wide fields or narrow fields. In your mouth the activated receptive field depends on how much of your food you allow to linger on your tongue! Vibration is not a customary tongue sensation whereas sensory nerves in the muscles of the tongue also detect tongue position in space. One difference from the skin is that with some odd exceptions, the tongue does not contain hairs! Werewolves might be the exception. There is a condition called black hairy tongue that is not really composed of hair follicles but overgrown tongue papillae. As a result of a chronic infection they enlarge and resemble hairs! It is also important to mention that "hot" foods, not referring to their temperature, do not actually involve taste. These activate TrpV1 vanilloid receptors of free nerve endings that we discussed earlier.

How then does taste work? Due to their cellular structure, taste buds are specialized to detect certain chemicals that help to decide determine whether the taste is acceptable. There are five basic types of taste sensations: sweet, umami, sour, salty and bitter. Some Type II taste buds cells detect sweet and "umami." Umami is also called "savory" as in broths or cooked meats. Monosodium glutamate is an umami taste. Within the taste buds and in taste cells are GPCRs that detect sugars and sugar alcohols. Did you know that "sweet" proteins also exist? I do not mean a barbecued steak that you might salivate over, unless you are vegetarian. These sweet tasting proteins are monellin and brazzein and they activate T1R2 and T1R3 GPCRs. T and R are simply ways to number different GPC receptors. Even more interesting is that these GPCR proteins for taste have a molecular structure like a venus flytrap! Along comes a sugar molecule or a sweet protein and zap (!), it is snagged by the GPCR protein. The protein then signals down the line that it has indeed captured a victim! Aspartame also gets caught by the protein flytrap, explaining why it is used as a sugar substitute. I will not discuss whether actual venus flytraps can taste the insects they consume. That might not be a pleasant thought.

Bitter taste is detected by T2R2 GPCRs in Type II cells. Interestingly these receptors can be fine tuned to very selective bitter tastes. One bitter taste is not the same as another despite their awful impact. Some bitter compounds activate several types of T2R receptors while others activate just one. Indeed, any given taste bud is thought to have 4–11 specific T2Rs that respond to specific bitter tastes. Thus a taste bud is equipped to handle many bitter tastes, each of which depends on its cells housing the right T2R2 GPCR receptor. Why should we be built to taste many varieties of unpleasant foods? I am referring to simply awful taste, not just a taste that a foody lists as less interesting. Might our world be better without tasting these molecules? It is possible that awful tastes might, like pain, help to protect us. This range of bitter taste receptor GPCRs is probably so diverse that its overall role is to protect humans from nature's toxins. What is one man's sweetness may be bitter to another. Roper and Chaudhari bring up the example of broccoli. Perhaps you thought that like or dislike of broccoli or its relative Brussels sprouts (brassica oleracea) [www. Widepedia] simply depended on your mother's reprimands. It's not so. Broccoli contains phenylthiocarbamide (PTC) and depending on your genetic makeup, it can taste terribly bitter, less bitter or even tasteless! You might recall that the late George H.W. Bush publicly commented that he hated broccoli. This of course, raised a storm of ire from protesting broccoli producers in the United States. It is not certain if this is why he lost the 1992 presidential re-election but it may be the only instance where T2R receptors were political. Your genetic code determines whether there are subtle changes in how your T2R bitter GPCR proteins interact with foods. If the food activates your exact type of taste receptor GPCR protein, you do not like it. If your GPCR protein is built slightly differently because of your genetic makeup, you might not even taste this. Of course, I am not sure a "foody" would approve of tasteless broccoli, but it might be a filler. It may also be that this kind of genetic variation influences whether you find brown beer bitter or whether you relish it.

Sour taste is next. A distinct mechanism is involved here. Sour seems to arise from acids. The cells that detect them are called Type III and they have acid sensing ion channels, ASCs, as discussed earlier. Apparently weak acids work best, like citric acid or vinegar (acetic acid). A unique story arises from ingestion of a West-African berry called Richadella dulcifica that contains a protein called "miraculin" (MCL). At neutral pH, MCL blocks sweet T1R2 and T1R3 GPCRs and offers no particular taste. However as pH is lowered, a direction that should generate a sour taste, MCL converts the sour taste into sweet! It's a "miracle." At lower pH, the 3D conformation of MCL changes, allowing it to instead activate sweet T1R2 and T1R3 GPCRs.[32] If you ingest a sour tasting food item, likely to be acidic, adding a berry to your tongue magically converts the sensation from sour to sweet. Exquisite biology is at play.

Salty tastes are a little more mysterious. They also apparently depend on other ion channels, not GPCRs. Too much salt is unpleasant, but it is unclear how this system informs you of it. Similarly it is not known whether fat, or fatty acids can generate specific taste receptors. Certain receptor molecules, including CD36 that is sensitive to long chain fatty acids, have been proposed to signal the sensation.[33] "Kokumi" is an odd taste that seems to feel like thickness. Once again, we do not know if this is a taste bud or taste cell signal or not. We also do not know if calcium is responsible for its own sensation.

So what happens after the taste cell, with its GPCR gets activated? What nerves and pathways are involved? Once the taste cells are activated by their GPCR receptors, they need to transmit the signal to the nerves that surround them. Like synaptic connections in the brain, specific neurochemicals are involved. Type II cells send ATP to their surrounding nerves. While you likely are aware of ATP as an "energy molecule," ATP is also a neurotransmitter. It has its own receptors called P2X receptors. Mice lacking P2X receptors become "taste blind" or perhaps "tasteless." This refers to their eating of something, not what the mouse itself might taste like, if referring to a cat.

These taste-blind mice cannot tell if items are sweet, salty, bitter or umami. You might wonder how a mouse tells us what they can taste? You can find out if you place electrodes that record nerve activity in response to, for example, sweet items. If an electrical response to the item does not appear when you are recording, the pathway is inactive. However, you can also reward mice with sweet foods and change their behavior. For example, you might design two connected cage spaces, one with tasty items and the other without. The mice will choose to spend much more time in the tasty cage. Did you ever wonder why fast food restaurants always have lineups? Tasteless mice, such as those lacking P2X receptors are not trainable in this way.

Besides ATP, there are also other neurotransmitters involved in taste buds such as our familiar Ach, this time from Type II cells. Interestingly, Type III cells release serotonin and GABA. You have heard of serotonin levels in the brain because antidepressant medications are used to keep their levels high. Serotonin in taste buds and one other neurotransmitter called GABA inhibit ATP release from the Type II cells. In this way, the Type III cells that release these inhibitory neurochemicals, weaken the taste signal from Type II cells. To make this even more complicated, some of the ATP from Type II cells activate Type III cells using another receptor called P2Y. Finally, it may be that some entire taste buds inhibit their neighbors. This is a property of other senses in the brain, such as the visual system. These are not discussed in this book, but the idea of "surround inhibition" related to vision led to a Nobel prize. The concept is that a group of cells respond maximally to a stimulus but inhibit their neighbors. This adds precision so that there is an exact "territory" of excitation, and not beyond the stimulation zone. In some ways, the tongue and its taste buds truly operate like a miniature brain with complex interplay between their parts. All of this means that taste system provides a "code" or "signature" that tells us what kind and combination of tastes occurred, how intense they were and where on the tongue they came from. The sensory nerves that respond to taste are "specialist," meaning they respond to one kind of taste, or "generalist," indicating that they respond to more than one. Both send signals that move upward to the brain. Here these sensations are further filtered, refined and connected to other sensations such as smell.

The PNS serves as the gateway for taste signals. In doing so, it refines the information sent to the brain. The brain then filters and puts these sensations into context. The nerve supply of the tongue itself is also not straight forward. The front two thirds including the tip is supplied for taste sensation by branches of the facial, or VII cranial nerve, including a branch called the chordae tympani. For other kinds of sensation, like pain, hot, cold or irritation, this front part of the tongue is supplied by the trigeminal, or Vth cranial nerve. The back, or posterior third of the tongue has both taste and other kinds of sensation supplied by the glossopharyngeal, or IXth cranial nerve. Although suggested in the past, we also know that tastes are not localized to some parts of the tongue only. Each part of the tongue is capable of detecting all types of taste. When I was a student we studied maps of the tongue that outlined portions sensitive to sweet, sour or bitter. Most of this is now debunked. This is not to say that much of what we studied in school was nonsense. Some parts were formative!

## Sensory nerves, not just bystanders

The bite of a mosquito tells us something interesting about sensory nerves. The saliva of the mosquito causes inflammation arising from a mixture of several proteins including an anticoagulant that prevents blood clotting and facilitates their feeding. The response by the host includes swelling, redness and heat from dilation of local blood vessels and leakiness of blood vessels. However, our ability to sense the irritation arising from a mosquito bite involves the activation of sensory axons in the skin. Unmyelinated sensory axons of the epidermis respond to inflammatory chemicals around them that include histamine, hydrogen ion (acid) and others called proteases. Recently investigators have also identified an "itch protein" in the spinal cord. Its proper name is NPPB (natriuretic polypeptide B).[34] It may be that NPPB magnifies itchy sensory input from the skin as afferents enter the spinal cord. However, here we focus on events at the nerve endings.

Beyond their ability to sense, nerves have yet another role. Sensory nerves also actively participate in inflammation and actually promote it. In legal terms, nerves are co-conspirators! Their involvement is termed "neurogenic inflammation," a two way interchange. On one side of the exchange, tissues signal axons with histamine, hydrogen ion and other mediators mentioned earlier. On the other side, sensory axon terminals locally release back into the tissue "neuropeptides," peptides that arise from neurons only, when they are activated. Peptides, in turn, are molecules that resemble proteins but smaller, consisting of only a few dozen amino acids instead of hundreds. Because neuropeptides are small, they can spread from nerve to nerve or nerves to other cells. Two neuropeptides we will discuss next are especially interesting and both have very specific receptors that respond to their presence.

Substance P (SP), mentioned earlier, is a small neuropeptide of only 11 amino acids but it is highly potent! It was discovered in 1931 by scientists von Euler and Gaddum at University College London during their studies of gut contractions. The "P" simply refers to "preparation" or "powder." This may sound a little unappealing but it was foundational. SP has a number of roles. Firstly it magnifies pain. Secondly it enhances inflammation by making neighboring blood vessels leaky. This is known as "plasma extravasation." Instead of blood or plasma remaining confined to small blood vessels, SP allows plasma to move out of the blood vessels into soft tissues. When tissues accumulate the leaked plasma, they swell, causing what is termed "edema." A small localized area of edema is also called a "welt." SP remains a topic of scientific interest. Small sized sensory neurons synthesize SP and as such, its presence is a marker of small unmyelinated "pain" fibers. Not all sensory axons contain SP and not all small sized pain fibers carry this neuropeptide. Why only some fibers should contain it and others not is a mystery. Perhaps its such a potent compound that only limited amounts are tolerable!

SP is not the only neuropeptide in sensory nerves. Others also participate in neurogenic inflammation. A co-conspirator peptide is CGRP, calcitonin gene-related peptide. This sounds complicated but is not. CGRP is mainly found in peripheral nerve cells and axons. It is actually more widespread than SP, present in about 40% of all sensory neurons. CGRP is a 37 amino peptide, a little larger than SP and comes in two flavors, α and β which may act identically but arise from different genes. Why this arrangement? Perhaps CGRP is simply important enough that evolution has assigned two independent genes to synthesize it. In any case, its gene also produces a completely unrelated peptide, known as calcitonin, a 32 amino acid calcium hormone produced by the Parafollicular cells of the thyroid gland, and not present in neurons. In essence then, two different peptides arise from the same gene and two genes produce the same peptide! I am not trying to confuse you!

Here is yet another example of the intense complexity of our biology. Let's focus on the two peptide one gene problem. We have two different peptide products that are made in completely different tissues-neurons or thyroid. This is known as "alternative splicing." You will recall that genes made of DNA, long strings of molecules called nucleotides, are codes for producing proteins and peptides in the body. Cells do this by copying the code from DNA to a related molecule called RNA. More specifically, it is called messenger RNA (mRNA), the middle step between the gene and protein. RNA then uses the cell to make proteins. However, genes often code for longer strings of DNA/mRNA than is strictly needed to produce a protein or peptide. To tailor the mRNA required for a specific protein additional splicing steps of the overly long mRNA complex are required. These steps are known as post transcriptional splicing since it occurs after DNA transcription into mRNA, but before proteins are synthesized. You may recall that splicing is something that film producers use, by cutting out questionable scenes to create a final version of the movie. Of course they may also splice out segments with particularly bad acting. What is spliced depends on the tissue and what its protein requirements are. For example, if the parafollicular thyroid cells specify one form of splicing, calcitonin mRNA and peptide are produced. Calcitonin is then delivered to the blood as a hormone. Conversely in neurons, the splicing decision is different and mRNA sequences are chosen to make CGRP mRNA and peptide. Specific proteins and enzymes in the nucleus are involved in splicing. Alternative splicing in summary means that the RNA generated by the DNA gene can be cut in different ways by different cells to generate unique peptides and proteins with very different roles. It's a remarkable process because it means that one gene has an expanded capability to synthesize several kinds of peptides or proteins. Its possible that film makers, unbeknownst to us, make several movies out of the same original film called the A version and the B version.

Like many neuronal peptides and proteins, CGRP was "under the radar" for many decades since its discovery in 1982. Recently, CGRP has proven to be of dramatic interest because if its role in migraine. It is one of the most potent dilators of blood vessels known. Given the discussion above about neurogenic inflammation, the role of CGRP, upon its release by axons, becomes obvious. It is responsible for dilated blood vessels in a zone of inflammation. CGRP also cooperates with SP in promoting pain. Many of the same small sensory neurons contain both SP and CGRP. Thus during local inflammation, the two neuropeptides are released together and collaborate. CGRP dilates the local blood vessels causing skin to redden and to warm. CGRP also has interesting roles during regeneration that will be discussed later.

Migraine headaches are form of neurogenic inflammation related to the release of neuropeptides, including CGRP. As one of the most common neurological conditions, its mechanism has been puzzling for many decades. CGRP is found in small sensory axons that supply the meninges, the connective tissue that surrounds the brain and that contributes to intracranial pain. It is found in close relationship to cerebral blood vessels that travel through the meninges and thereby it promotes pain and dilates blood vessels. CGRP may thus account for key symptoms of migraine: a pounding painful headache and rises in blood flow. Several pharmaceutical manufacturers have developed anti-CGRP therapies with initial clinical trials showing striking benefits in migraine. However, given its importance in the PNS and sensory nerves, one wonders of the long term impact of inhibiting CGRP in this way might be.

"Neurogenic inflammation" refers to an expansion of an area of inflammation because of the participation of nerves. Experimentally cutting skin nerves reduces the extent of skin inflammation. One important theory is that branches of nearby sensory axons are activated first as any local form of inflammation begins. As these nerves relay information upward toward the brain, they also activate other branches, found within the adjacent skin, of the same parent nerve. When one sensory branch is activated, a sister branch of the same parent axon might thereby participate (Fig. 21). Since nerves actively participate in inflammation, these branches act to enlarge the area of inflammation by their release of SP, CGRP and other constituents into the skin. Targets for the neuropeptides, beyond blood vessels include "mast cells," specialized cells found in most body tissues that are packed

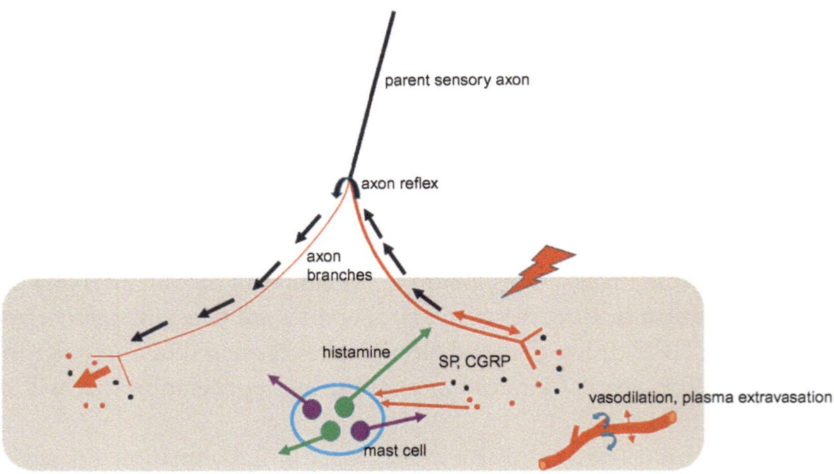


**Fig. 21** Diagram illustrating neurogenic inflammation. A stimulus activates a sensory axon branch sending a signal to the branch point. At the branch point, there is antidromic (reverse direction) signaling in a second axon branch. The sensory branches release the peptides SP and CGRP that dilate vessels, induce plasma extravasation and degranulate mast cells. Mast cells release histamine and other constituents that reactivate axons.

with inflammatory molecules including histamine and serotonin. As this sequence goes, activated nerve endings release neuropeptides that activate mast cells that in turn release histamine. Histamine has long been linked with allergy and inflammation and causes itch, perhaps in collaboration with NPPB in the spinal cord! Histamine may also activate additional nerve fiber endings thereby enlarging the zone of activation and inflammation. In summary there is a loop: nerve terminal-neuropeptide-mast cell-histamine-nerve terminal. This loop, hypothesized a number of years ago is called the "axon reflex." While oversimplified here, it does explain how nerves can be culpable in inflammation and do not always help the situation! More importantly, they let us know of a problem in the making so that we might try to alleviate an uncomfortable area of itching and redness from an unavoidable mosquito bite. Of course scratching an area of inflamed skin is also a certain way to enlarge inflammation!

# CHAPTER 3

# Irreversible events: How nerves are injured

## Why don't we hear about these?

Injuries to nerves are the most common neurological injury. You may not realize that head and spinal cord injuries are actually much less common than nerve injuries. Why is this fact so little appreciated? This is because nerve injuries are dealt with day and night in emergency rooms of hospitals across the globe. There is little fuss or epidemiology done over common nerve injuries; the patients' wounds are resutured either immediately, or a few days later by a plastic or hand surgeon. But be aware-this does not imply that the problem has been solved, as will be discussed below. Nerves can be resewn surgically but they frequently do not fully recover. One of my plastic surgery colleagues had noticed that certain times of the year were busier for this sort of work than others. The busiest were over the Canadian Thanksgiving weekend, a time of family celebrations and get togethers. It is also a time to carve a roasted turkey with sharp knives. Both inexperienced and experienced "carvers" visit hospital emergency rooms with lacerated or severed fingers! In violent places where knives are weapons, a variety of other kinds of nerve lacerations can occur.

Nerve injuries are also a feature of multitrauma scenarios in which the damage is extensive and involves several organs. Here, the initial focus may be upon fractured long bones and pelvic injuries, admittedly serious medical emergencies. The associated nerve injuries, often not immediately recognized, may cause persistent difficulties long after casts has been removed. Since nerve injuries are not life threatening, they usually receive less attention. In a patient with a chest tube who is unconscious from a head injury and has multiple long bone fractures, detecting nerve injuries can be challenging. Limbs may be swollen from fractures, surgery or soft tissue injuries. Sometimes nerves are damaged from "compartment syndromes." These occur because swollen damaged muscles and other soft tissue injuries are confined in specific parts of the limb. These compartments, such as the front of the leg below the knee (anterior tibial compartment) have

*Our Wired Nerves*
https://doi.org/10.1016/B978-0-12-821487-9.00003-9

rigid tissue barriers around them called fascia. Normally there is plenty of room in a given compartment but if there is an injury, the contents swell or may have accumulated blood. If the swelling outgrows the available space in the compartment, it compresses the blood supply. An interrupted supply of blood for six hours or more causes muscle cell death, known as "necrosis." What is less appreciated is that within this swollen tightened compartment, are key nerves that travel to the foot such as the deep peroneal nerve (now called fibular nerve by anatomists!). As an aside, "ischemia" refers to loss of the blood supply to a tissue (strokes are secondary to brain ischemia; "heart attacks" are from cardiac muscle ischemia). Compartment ischemia lasting more than 3 h permanently damages peripheral nerves. Damage to the deep peroneal nerve causes a foot drop and numbness in the foot. The fate of important peripheral nerves is at stake in multiple injuries complicated by ischemic compartment syndromes. If surgeons are required to remove or resect traumatized tissue adjacent nerves may be damaged by the operation. It may be challenging to identify the nerve in a bed of swollen traumatized tissue!

Some other nerve injuries are not difficult to identify. My most memorable was of a cyclist who was attacked by a bear near Banff Alberta. For unclear reasons, the bear thought the man's shoulder was worth biting and tearing apart. Luckily for the person, his injuries, albeit severe, were confined to the shoulder. However, the bear managed to resect several inches of the brachial plexus, the complex network of nerves that travels from the spinal cord in the neck, through the shoulder and into the arm, forearm and hand. In addition to loss of the muscles and soft tissue of his shoulder, he also lost the entire nerve supply to his arm, forearm and hand. This rendered complete paralysis, loss of sensation and ongoing pain. For injuries this severe, not only is the extensive lost portions of nerve impossible to replace or bridge but its normal environment within the shoulder is severely disrupted. In the face of such a serious scenario, the immediate attention requires prevention of exsanguination (severe blood loss) from his nearby axillary artery. Ensuring that all of the neighboring nerves have an adequate blood supply at the same time is probably beyond the capability of modern trauma surgery.

## What kinds of nerve injuries are there?

Nerves are actually surprisingly difficult to injure. Neither mechanical nor "ischemic" (interrupting its blood supply, see above) injuries that damage nerves are trivial. Sufficient force, or laceration of tissue is required. Ischemic injuries require that a number of feeding blood vessels are interrupted. Nerve surgeons classify nerve injuries into several categories that

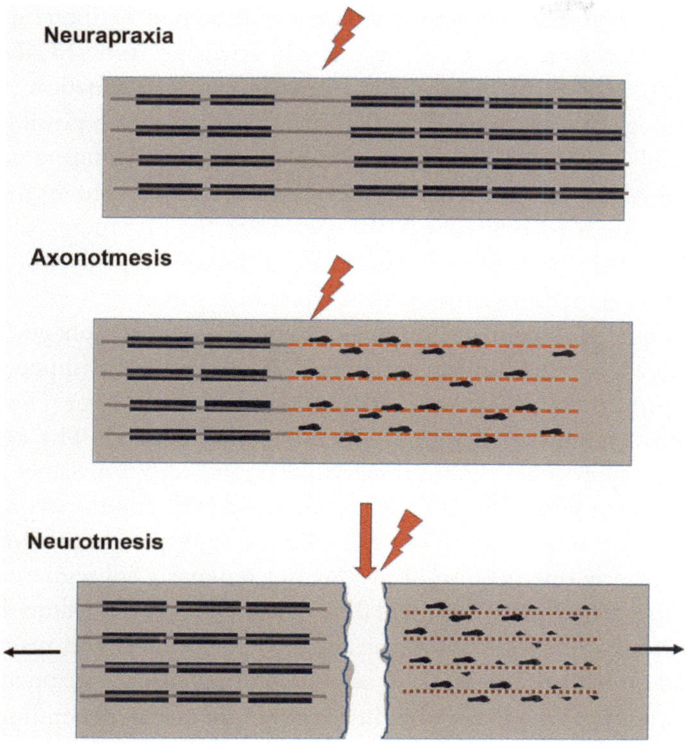


**Fig. 22** Diagram illustrating Seddon classification of nerve injuries: demyelinating neurapraxia, axonotmesis with axonal degeneration and neurotmesis with nerve trunk disruption and axonal degeneration.

relate to the prognosis for recovery. Milder injuries are categorized as those that interrupt the myelin only. More severe injuries damage the axon and the most severe injuries not only damage the axon but also physically separate the nerve apart into two stumps. These are the proximal stump, closer to the spinal cord and the distal stump on the other side of the separation. In 1942, Herbert Seddon (1903–1977), an Oxford orthopedic nerve surgeon, classified types of nerve injury in a way that remains highly useful to our understanding. The types of nerve injury are known as neurapraxic, axonotmetic and neurotmetic injuries (Fig. 22).

## If a nerve injury occurs, this is the type you might prefer

The most favorable types of nerve injuries are called neurapraxic lesions. Neurapraxic injuries disrupt the myelin alone sparing the axon within, “demyelinating” injuries. They follow from blunt trauma or compression damage. Neurapraxia requires demyelination sufficient to interrupt transmission, a consequence of disrupting the delicate axon-SC-Node of Ranvier. There is some debate about how much demyelination needs to occur to block axon transmission. For example, it might be possible for an AP to "skip" over a single segment of myelin loss. Loss spanning two consecutive segments of myelin (two "beads" or sidewalk cement segments, see above) is enough to interrupt APs experimentally. In patients disruption of the node may be sufficient. Subtle distortions of the myelin, through inflammation or trauma is not well tolerated by parent SCs who respond by degrading their own myelin! Soon neighboring macrophages, the tissue scavengers of the body, also participate. The myelin is stripped from its segment and the SC then considers synthesizing a new layer. The process of myelin stripping by macrophages is intriguing to watch. The aggressive macrophage approaches the myelinated nerve and inappropriately inserts a part of its cell in between the myelin lamellae (layers) and the axon, gradually separating the layers away. The SCs and macrophages are not content to simply take away one portion of myelin, but remove a full segment. This is why pathologists identify "segmental demyelination" as the hallmark lesion. If the next adjacent segment is normal, its Schwann cell and macrophage may decide not to touch it. Thus, segments of myelin are stripped, sometimes apparently at random along the length of the nerve, depending on the type of injury or disease. Another type of demyelination simply alters the architecture of the node. "Paranodal widening" alone may be enough to interrupt signal transmission.

What happens to nerves that are demyelinated? Several properties of myelinated axons are changed by demyelination. Since intact axon architecture is crucial to the operation of ion channels, segmental demyelination alters their properties. The channels may, for example, spread along the axon beyond the Nodes of Ranvier, fail to cluster properly and not consolidate their depolarization. They may also be destroyed by the problem that damaged the myelin, such as an inappropriate exposure to inflammation. Demyelination in general is a problem because it delays the time required for axon currents to reach threshold to form an AP. This slows current flow and conduction "speed" or velocity is reduced, an important hallmark of demyelination. A more serious consequence of demyelination is complete interruption of conduction. In this case, the ion channel current flow is not sufficient to generate an AP. When the action potentials fail to jump from node to node, information stops and signals do not move along the axon. This process is known as "conduction block."

To determine if a person with a neuropathy has demyelination, their electrical conduction can be studied. Described in more detail later, electrophysiology studies to sort this out are called "nerve conduction studies." They are also simply called EMG studies. This term, however, is a misnomer because EMG is short for "electromyography" or measurement of the electrical properties of muscles. EMG involves using needle electrodes that are directly placed in muscles to record muscle activity. Nerve conduction studies involve stimulating and recording from nerves and they inform us about conduction velocity among other details. In practice, both nerve conduction and EMG techniques are regularly used to study patients with nerve injuries. Together they are called electrophysiological or neurophysiological studies but we will discuss them in more detail in a later chapter.

If a nerve has demyelination, its conduction velocity is slowed, as described above. Slowing can be quite severe from a normal conduction velocity of 50 m/s down to 10 m/s or less. However, most of us would not notice if our conduction velocities were altered. This is because even 10 m/s far exceeds our reaction and action times. What we would notice however, is conduction block, failure of the signal to transmit at all. In this case the signal does not transmit through the demyelinated zone, and it is impossible to activate a muscle, or to have a sensory stimulus arrive at the brain. Given this, demyelinating neuropathies can be devastating, causing complete loss of function. If they involve most of the nerves of the body, the paralysis can be extreme. The only good news side of this story is that demyelination can recover without the axon having to regenerate. Thus, although myelin is stripped off of the axon, the remaining axon can be remyelinated by a new or a repentant Schwann cell. Hence a demyelinating injury is preferred if you were destined to sustain a nerve injury.

Myelinated axons experience a unique form of damage from compression. Their delicate architecture and nodal structure are disrupted by injury more easily than unmyelinated axons. As the nerve is compressed, one internodal segment of the nerve is pushed ahead and underneath the next internode, like a retractable telescope lens. This is also called intussusception, a term used for telescoping of the bowel walls in young children, a cause of acute and serious abdominal pain. When this telescoping happens to the nerve, both Schwann cells and macrophages detect it. They then initiate their programs that call for destruction and dissolution of the myelin segments that were distorted by the intussusception. The result is demyelination of one more segments.

## Nerve damage that does not recover well

"Axonotmesis" injuries represent the next category of disruption to nerves. These are lesions that destroy a part of the axon while leaving the general architecture of the whole nerve trunk intact. Put simply, the wiring is disrupted but the overall nerve connective tissue remains undamaged. Axonotmesis injuries arise from nonpenetrating injuries, or blunt trauma that is intense enough to distort and damage the axon. Severe stretch injuries might accomplish this. Why are nerves susceptible to this kind of injury? Could not the axon simply reconnect, reform and recover?

When sufficient injury to the axon occurs, an irreversible damage signal is sent along it. What exactly triggers the injury signal is uncertain, but it involves not only the site of axon damage but outward along the entire axon tree. This is a self-destruct signal that triggers eventual breakdown of the axon beyond the injury, but spares the proximal axon above the injury site. The process is called axonal degeneration (AxD). If axons are actually cut or transected, the term "Wallerian" degeneration is used, named after an anatomist Augustus Waller (1816–1870), a British neurophysiologist and anatomist. An interesting note is that he died from coronary artery disease at age 53, but was survived by a son of the same name, also a physiologist who invented the electrocardiogram (https://en.wikipedia.org/wiki/Augustus_Volney_Waller)! Sometimes family members are motivated by personal tragedies and challenges.

When a complete transection of the nerve has not occurred, stretch or distortion of the axon membrane is the trigger for AxD. The damage signal that initiates degeneration is thought to be initiated from an excessive inflow of calcium into the axon. Small amounts of calcium shuttle through axons normally, interact with proteins as part of their normal physiology. If they are of a sufficient degree and magnitude however, higher calcium levels can trigger a damage response, AxD. This self destruct sequence involves an extensive and orchestrated series of events. For example, a "cascade" of proteins known as caspases and calpains eventually digest most of the other proteins within the axon. This in turn triggers neighboring Schwann cells and macrophages to consume or "phagocytose" the axon, its contents and its myelin sheath. The disrupted materials, such as pieces of axons or myelin, are eventually cleared and disappear. Think of those alarming time lapse videos of the seabed, in which a dead fish is rapidly consumed by local residents like crabs and worms! After a short period, nothing is left! The full destruction sequence of an axon involves a series of protein interactions from beginning to end. Of course, despite the intricate biology, the irreversible fate of the axon is not elegant for the victim.

Injury signals do not cause AxD of proximal axons above the injury site that remain connected to the cell body. While some limited "retrograde" damage develops to proximal axons, it is limited to one or two myelin segments only. The proximal axons become the parents of new regenerating axon sprouts, discussed later. Degeneration of an axon beyond the injury site was once thought to occur because of its simple separation from the cell body. Since proteins can no longer be transported past the injury site, it was thought that degeneration arose from deprivation, i.e., the axon beyond the injury site is "starved" of essential material. This thinking was challenged by an interesting set of findings involving a mouse strain with nerves that apparently refuse to die! These unique mice, called $WLD^s$ do eventually develop AxD but its process is very slow and delayed. In other words, when the axons were cut, the stump at the far end of the transection, also known as the "distal stump" simply did not undergo normal AxD. One would normally expect terrible carnage of distal axons within days of a cut. The axons would have shrunken or disappeared, the myelin appears fragmented into balls called "ovoids" (yes they are roundish!). The SCs become "activated," multiplying, migrating and changing their proteins. The $WLD^s$ mice did not readily develop these changes, a discovery that unearthed a series of new AxD proteins. As a result, we now recognize that axonal degeneration is an active and sophisticated process, not a passive result of deprivation. The newly identified molecules, potential targets for new treatments, offer hope that unwanted AxD may be prevented.

The biochemistry of AxD gets complicated, but bear with me. Mutant $WLD^s$ mice were found to overproduce am axon molecule called Nicotinamide adenine dinucleotide ($NAD^+$). $NAD^+$, in turn, is critical for handling energy metabolism in axons and its loss causes AxD. It is produced, in turn, by an enzyme called Nicotinamide mononucleotide adenylyltransferase (NMNAT). Our mutant $WLD^s$ mouse had overactive NMNAT allowing its axons to sustain robust $NAD^+$ levels during AxD, thereby slowing its progression. Through the work of Drs. Jeff Mildbrant, DiAntonio from the Washington School of Medicine and others, we now know that this "protective" axon pathway includes a series of additional molecules called SARM1, DLK, JNK and others.[35] How they all fit together and what other molecules are also involved is still under active investigation, including some by our own laboratory.[36] One bottom line is that this series of events happens before the calpains and caspases, both apparently aggressive molecules, begin to digest the axon proteins. Surely this is very significant. Since $WLD^s$ mice have very slowed AxD, subsequent regeneration is delayed, an outcome that may not be an advantage. However, we know from discussing neuropathies that AxD can also occur from other types of injuries besides cuts and crushes. Chemotherapeutic medications, lack of thiamine, inflammation and others trigger AxD to cause neuropathy. Since the AxD from these diseases is identical to that following injury, slowing or preventing AxD would be a very valuable new treatment approach to consider.

As of today, treatments to arrest the axon injury signal and to prevent degeneration do not exist. Once this complex process is triggered and set in motion immediately after injury, it is irreversible. Similarly, axons that have been separated cannot be reconnected. AxD rapidly follows. Recovery from axonotmesis requires a series of steps. None are very satisfactory because they are slow and often incomplete. Firstly, axonal damage has to run its course. Axon and myelin debris must be cleared away, since they can inhibit new growth. SCs next must prepare guidance pathways for new axons to grow along, an essential role in directing regrowth. New axons sprout from parent axons in the stump proximal to the injury. They then traverse the injury zone into new pathways formed where previous axons have died. New axons elongate at 1–3 mm per day at best. For a 500 mm limb such as a short leg, growth might require 6 months to 2 years under ideal circumstances! Thus recovery requires slow and protracted regrowth of axons back to their previous targets where they belong. As will be discussed later, the process is highly limited and often unsuccessful.

Is the target tissue, previously innervated with axons, waiting for axons to return? They do not wait forever. Slow axon regrowth is probably the greatest liability to recovery after a nerve injury. When regrowing axons eventually reach their target destinations they may no longer be warmly received. For example, even as the slowly growing axon moves farther along the distal nerve stump but has not yet reached the target, it may eventually feel unwelcome and may wither. The SCs guiding them are meant to prevent the axons from giving up on their quest and are there to support them. However, SCs also may eventually give up the ghost. Target muscles may shrink (atrophy) and harden into inexcitable connective tissue. Even before this happens, they may not have proper endplates left, with the correct configuration of molecules and receptors for the motor nerves to connect to. The exquisite sensory organs we have described simply disappear over time. Overall, the progress of recovery from axonotmesis is significantly impaired. The belief that "peripheral nerves regenerate robustly" has serious limitations, usually arising as a mistaken attitude of investigators working on the brain, spinal cord or other areas. In reality the situation is very different. There is no surgery or treatment that speeds recovery in victims of this type of nerve damage.

## Nerve damage that rarely recovers

The most serious category of nerve damage is neurotmesis. This injury not only transects the axon, but it cleaves the nerve trunk into proximal and distal stumps. These nerves are truly torn asunder. The axons, perineurium and epineurial sheaths are divided and their recovery requires not only regrowth of the axon but reconstitution of its coverings. Nerve trunks are normally under elastic tension because they include a protein called, for predicable reasons, elastin. Elastin is present in the nerve sheath and its role is to keep nerves and tissues nicely aligned. The impact of elastin on the nerve is interesting to observe by microscopy. When a nerve is removed along its length and laid out, individual axons within the endoneurium can be appreciated and they appear wavy, like the folds in a drawnback curtain. When the nerve is stretched during normal movement, axons slowly unfold, preventing them from individually being torn and damaged. As stated earlier, it is essential that the delicate architecture of the nerve, especially at Nodes of Ranvier, not be disrupted. After a neurotmesis injury, when the nerve trunk and its elastin are transected, the distal and proximal stumps retract or pull away from one another like a cut elastic band. This understandably leaves a new gap. Sometimes a talented surgeon can gently pull these stumps back together into position and suture them. However, if too much time has elapsed or the gap is too large, this is not possible. Our unfortunate cyclist with an extensive brachial plexus resection, courtesy of a Banff grizzly bear, could not be treated this way.

Axons must bridge or jump this gap to recover. However, they prefer not to. Not only are the axons reluctant to traverse the damage zone, but they have little guidance to help them make the jump. When successful jumping axons eventually reach the other side of the gap, they must next discern which SCs they should associate with. As axons reach the other side, many SCs are available and waiting, but are often strangers without any previous relationship with the axon approaching them. Overall, given all of this choice, it seems unlikely for a given axon to identify the correct pathway to its target. By "correct," an axon sprout from a parent axon should be directed into the pathway it occupied prior to injury. This would ensure the axon finds its original target and performs its original set of duties. Whether an axon can identify the distal stump at all is also debatable! However, help is available, since thoughtful surviving SCs from the distal stump migrate backward in search of the proximal site. SCs coming from the distal stump enter the gap zone where they likely send out attractive cues. Recall that in the distal stump, the axons undergo degeneration but SCs survive and live for another day. In fact, these SCs, no longer under a relationship with viable axons, change their behavior, becoming mobile and promiscuous. By traveling backward they welcome newly sprouting axons, at least in theory. However, axons may veer away and miss both the signals and the distal stump altogether. For example, axons sometimes double back and enter the nerve sheath of the proximal stump in error. When bundles of axons grow inappropriately without reaching the distal stump they often form structures called "neuromas." These are enlarged, sometimes painful collections of connective tissue and misdirected axons.

This overall failure of cut nerves to navigate their way home was beautifully illustrated in a paper by Witzel, Brushart and colleagues[37]. Using a genetic mouse that has individual fluorescent axons, these investigators imaged nerves attempting to regrow after transection. Their images show wild and random attempts to jump the injury gap zone, often unsuccessfully. To make the point, Brushart's textbook, "Nerve Repair"[38] illustrates the problem on its cover. Estimates are that only one in ten axons bridging transections successfully reach their targets. This is a dismal outcome and happens despite careful microsurgical reconnection of proximal and distal stumps. For stumps that cannot be surgically re-apposed, biodegradable tubes or conduits or tissue grafts have been placed across gaps. These can offer some degree of guidance. Given all of these issues, it is clear that neurotmesis nerve lesions are the most severe category of nerve injury. Despite careful microsuturing or other approaches, they have poor recovery and again challenge the comment that "peripheral nerve regenerate robustly"!

Injuries can be surprisingly complex with issues beyond the three categories we have just described. There are also more detailed classifications such as that by Dr. Sydney Sunderland (1910–1993), an Australian physician, neuroanatomist and former Dean of the University of Melbourne, who divided axonotmesis into different levels of severity. For example, there may be partial disruptions of the fascicles of the nerve but the epineurial connective sheath remains intact. Moreover, many injuries can include degrees of both of demyelination and AxD (Fig. 23). The severity of the injury can depend on the proportion of axons directly damaged and that develop AxD. The long term prognosis and recovery is largely determined by the degree

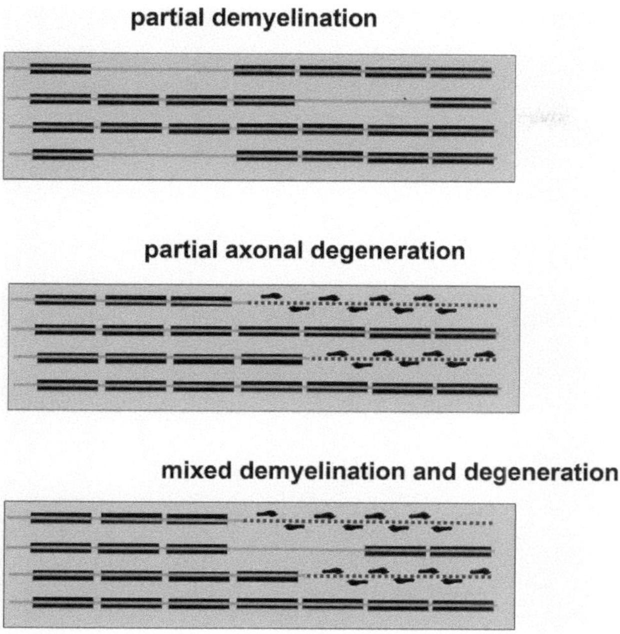


**Fig. 23** Diagrams illustrating common partial nerve lesions that include demyelination, axonal degeneration or a combination.

of axon damage rather than demyelination. Of course complex real world injuries such as the bear attack injury we described earlier escape classification and fall into their own category. There is no doubt that nerve injuries are challenging!

A final interesting concept is what has been called the "bystander effect." This refers to damage in nerves that spreads beyond the targets of that original injury. An example of this can be found following nerve inflammation, as might occur in an autoimmune disorder. While the primary immune targets may be a myelin protein, the associated and intense inflammation can result in damage and AxD of nearby axons. In this case, although the disorder may be thought of as "demyelinating," outcomes are more severe because neighboring axons are irreversibly damaged.

# CHAPTER 4

# Are there nerves? How to test the peripheral nervous system

Some have espoused the view that modern imaging techniques such as MRI, ultrasound and others might yield a complete "picture" of the peripheral nervous system. This would be exciting and has been summarized by Dr. Guido Stoll from Wurzburg Germany, a leader in this work.[39] However, the resolution for such detailed images is not yet possible. One challenge is the distortion of small nerve images by nearby muscles, bones and blood vessels. Injecting a substance that is taken up by nerves and is easily seen by the MR magnet might be an alternative approach and is being tested using iron particles.[39] An additional consideration is that the peripheral nervous system extends to the farthest reaches of the body with end branches separated only by microns. Dramatic upgrades in resolution would be required to view this entire "tree." Fine terminal branches would overlap so densely that they would outline every wrinkle and twist of the skin in high detail. Given the current growth in capacity to store vast inputs of data, described as "big data," it is conceivable that an individual human map of an individual's PNS could be constructed. A low power view might image the entire body or a limb, making it obvious if one segment were missing or altered as a result of damage. Individual branches might be detected with yet higher power imaging with greater scrutiny; early and subtle changes in a polyneuropathy might be detected. For instance, if sensory axons are retracted from the ends of the extremities in very early disease, this might be visualized. Clinical trials might be based on following the regrowth of nerves using advanced imaging. These are interesting and hopeful possibilities not yet realized.

## The neurological examination: Honed classicism

Since advanced "nerve scans" are not yet available, we continue to rely on classical "tried and true" examinations of the peripheral nervous system in humans. An additional point is that even the most exquisite anatomical map of an individual's nerves would not necessarily tell us if they are working properly. To infer function from subtle changes in nerve size or structure

*Our Wired Nerves*
https://doi.org/10.1016/B978-0-12-821487-9.00004-0

will require another revolution in data capture beyond the whole body maps envisioned above.

Over the last two centuries, neurologists have developed and honed an elegant and highly informative method of examining peripheral nerves. It is called the neurological examination! By the early 1940s–1950s the classical neurological examination was a well defined art. Designed to assess all disorders of the nervous system, it has been partly displaced over the last 30 years by brain and spinal cord MR (magnetic resonance) and CT (computerized tomography) imaging. With the advent of scans, neurologists are probably less diligent and skilled in routinely performing this examination. Other physicians live in fear of it! It is complex and can be time consuming. Significant experience is essential in using the neurological examination properly. The role of the neurologists has been interpreted as performing neurological examinations and localizing where in the nervous system the lesion was, now a misplaced bias from previous times! Rounding out this picture was an image of an austere white coated male with a bow tie and pins in the lapel for checking sensation! Times and neurologists have changed and pins are out (along with bow ties for most neurologists). Stroke neurology for example has become an action packed surgical-like specialty. A little of the older honed classicism is crucial during examinations for peripheral nerve disease and remains the gold standard for assessing nerves and muscles. A combination of carefully chosen scans and detailed examination will detect the disorders of nerves discussed further in this book.

The neurological examination tests for motor, sensory and reflex changes in the head and neck, limbs and trunk. It begins with an assessment of the person's mental status and language. Then it evaluates each cranial nerve. Next is the motor system. Abnormal movements like tremors are observed. The muscles are checked for loss in size, known as atrophy. Muscles are sometimes observed to have subtle ongoing involuntary twitching called fasciculations that we discussed earlier. A grading scale known as the MRC (Medical Research Council of the UK) classifies the degree of weakness for all voluntary motor movements. Thus 0 is complete paralysis, 1 is a twitch of movement, 2 is movement but not enough to overcome gravity, 3 is just enough strength to overcome gravity but not against added resistance, 4 is strength against resistance but not full strength and 5 is normal.[40] Some argue that the MRC scale should be replaced because it is not linear and that 3 or 4/5 have wide variations. Despite this, other scales have not improved this simple standardized scale, useful when comparing examinations by the same examiner from visit to visit. The Mayo Clinic has an alternative scale but it is not widely adopted and has some of the same challenges. It is important to recognize that an MRC "5," or normal, for an 89 year old woman is not a "5" for a 20 year old kickboxer (a neurologically imprudent sport). Age and muscular size-grading is a judgment call on the part of the examiner. A 5 simply reflects appropriate normal power for a person of a given size or age. With experience, muscles are easy to grade. Next are the deep tendon or stretch reflexes that we discussed in detail above. Common deep tendon reflexes tested include the biceps that flexes the arm, the brachioradialis that also flexes the arm but with hands held vertically (thumbs up), the triceps that extends the arm, the knee or quadriceps reflex that extends the knee and finally the ankle or Achilles reflex that plantar flexes (bends down) the foot. To test sensation, examiners use light touch (cotton swab or brush), pinprick (clean nonreused unclasped safety pin-not penetrating the skin!), vibration (128 Hz tuning fork) and position sensation (identifying whether a finger or toe was moved up or down with eyes closed). Testing thermal sensation may involve using tubes of hot and cold water. A simpler test of cold sensation is to simply run cold water over the ends of a tuning fork and testing how sensitive the subject is to cool metal. Of course in Edmonton, the difficulty is that the extremities can already be very cold in winter, so the added cold metal of the tuning fork in the toes may not seem like a cold stimulus! Next, it is essential to watching a person walk. For example peroneal neuropathy causes a foot drop, an inability to raise the toes, that causes the foot to slap down when walking. Experienced neurologists can hear these people walking down the hall! Having a person stand motionless with arms outstretched, palms up and eyes closed can test for position or proprioceptive sensation. If the person loses step and is unable to remain standing steady it may be because they are unaware of where their feet are! This is known as a Romberg sign, named after Moritz Heinrich Romberg (1795–1873) [https://en.wikipedia.org/wiki/Moritz_Heinrich_Romberg], a neurologist from Berlin who studied tabes dorsalis, a neurological complication of syphilis! In the relatively few clinical research trials designed for new polyneuropathy treatments, some of the key elements of the neurological examination are graded using a simple score. These have geographical names like the Utah, Toronto or Michigan neuropathy scale and others.

## Diagnostic shocks and needles

We have compared nerve axons to hollow centered cables. Since an electrician is required to test how well her or his work is connected, a similar approach is adopted for nerves. In the same manner that an electrician tests for electrical connectivity, the study of "electrophysiology" examines the "physiology" of the nervous system. Another term is "neurophysiology." Physicians and investigators have spent almost a century performing various types of electrophysiological testing but its current form has steadily evolved over the past 50 years or so from pioneers including John (Iain) Alexander Simpson from Scotland, Fritz Buchthal from Denmark, Edward Lambert from the United States and others. Many of the tests are standardized and competency is maintained in most jurisdictions by a qualifying examination. While a large number of interesting variations on nerve electrophysiology testing have been published, it is the standardized tests that are most likely to help patients and categorize their diagnosis. The physicians that perform these tests are either neurologists or physical medicine specialists. In some countries, specialized "neurophysiologists" have been employed but without rounded neurological and medical training they are at a disadvantage. One of my neurophysiology mentors, Dr. William Brown, routinely reminded us that neurophysiology is "an extension of the neurological examination".

Most modern neurophysiological (electrophysiological) examinations consist of two main parts. Nerve conduction studies are recordings from motor and sensory nerves. EMG (electromyography) involves the insertion of fine sterilized nonreusable needles into muscles. For nerve conduction studies, stimulation of the nerves at well established landmark sites activates the nerve through the skin. Standardized stimulation sites indicate nearby nerves, accessible to an electrical current traveling through skin to activate them. Conductive electrode paste is used to facilitate the stimulation. Good electrophysiological laboratories remember to wipe this messy paste off of the subject when the study is done! Once the motor axons are activated, electrical activity is measured and recorded by a recording electrode positioned over the relevant muscle. Responses from activated motor axons are called CMAPs, "compound muscle action potentials" and they are recorded at the same time the muscle twitches. Activation and recording from single motor units and axons is challenging. In the laboratory, enough current is applied to instead activate all of the motor axons and motor units within the nerve being stimulated. In this way, the maximum amount of muscle activation can be measured. A 20% further increase in stimulation intensity is usually provided to ensure all of the motor axons have been activated and tested. Thus, the CMAP (called an M wave in some places) is the summation of many individual motor axons connected to many individual muscle fibers. This is where the "C" in CMAP arises, i.e., a compound response from many motor units. When the recording electrodes are properly placed, they detect signal from all of the muscle fiber action potentials near the NMJ (the endplate) summed together.

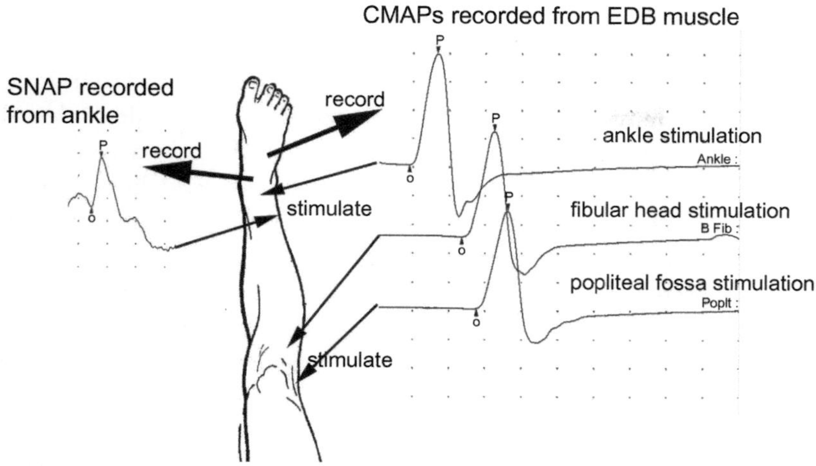


**Fig. 24** Examples of recordings of normal human nerve conduction. This is a study of the peroneal nerve in the leg. The motor peroneal nerve is stimulated at the knee, fibular head or ankle and compound muscle action potentials (CMAPs) are recorded from the EDB (extensor digitorum brevis) muscle over the top of the foot. Note the large graceful appearance of the CMAPs that appear similar irrespective of where the nerve is stimulated. A sensory nerve action potential (SNAP) from sensory axons of this nerve is recorded following stimulation near the ankle and recording over the nerve on top of the foot. SNAPs are much smaller than CMAPs.

CMAPs are graceful traces to watch (Fig. 24). They form a broad smooth upward stroke that appears after a short interval known as the latency of the response. This simply refers to the time it takes from the stimulus when given, to the onset of the response, a latency. The trace then returns toward baseline, overshoots below it and finally flattens back to quiescence. Some of us never lose our fascination over these traces. Of course, when they are abnormal it is a different, equally fascinating story.

Many pieces of information come from recording CMAPs. The size or height of the CMAP (called amplitude) depends on the number of motor axons that are viable and can be recruited, or stimulated. It also depends on the numbers of muscle fibers at the recording sites. Thus a loss of the CMAP might arise from loss of motor axons, from motor axons that are present but do not conduct properly or from signal that fails to cross the NMJ. A disorder targeting muscle fibers (known as a myopathy) may have small CMAPs because muscle fibers have disappeared. After the stimulation, the slight latency or time delay preceding the CMAP, is officially denoted as the distal motor latency, also called DML. If the DML is abnormally delayed, it suggests that either Nodes of Ranvier, or myelin are abnormal. The intelligent and diligent reader will then be skeptical. How can a simple time be informative? To begin with, some hands can be very tiny or very large and DMLs would be greater as they travel further in a large hand! The distance from the nerve stimulation site to the recording site is expected to be longer in our kickboxer with big hands than our 80 year old female (unless she happened to be unusually tall). As it turns out, DMLs have a standardized normal range for a variety of hand sizes. Standardized guidelines provide a range of DMLs to the examiner that are linked to recommended distances between stimulation and recording. These normal values are essential and indicate when the response is outside what might be expected, even in the case of variable hand sizes! Hands and fingers are larger in most men than women, but values fall within an expected range. By stimulating the nerve at two different sites along the arm or leg, a conduction velocity (CV) can be calculated since the distance is known between these sites, and the time required can be calculated. CV is the distance divided by the time taken. The CMAP normally has a similar appearance independently of where it is stimulated, because it represents a summation of all of the motor units of the nerve, irrespective of stimulation site. Of course, changes in the CMAP between sites may indicate that the nerve between them is abnormal, perhaps demyelinated. CMAPs can be fully or partially blocked between two different sites or their wave shape can spread out because of demyelination, a change known as dispersion.

Nerve conduction studies also examine sensory nerves. In this case the sensory nerves are stimulated through the skin, and their summed action potentials are recorded either further down the limb (antidromic conduction, the opposite direction of normal sensory conduction) or further up the limb (orthodromic, or in the normal direction of sensory axon transmission). As we have discussed earlier, stimulating and recording from nerves is artificial and nerves have identical conduction velocities whether the technique is antidromic or orthodromic. The response from sensory nerves is called a SNAP, which stands for sensory nerve action potential (Fig. 24). Like CMAPs, they are a summation of APs from all of the single sensory axons that are being recorded. However unlike CMAPs, the recordings are from nerves, not muscles and SNAPs are considerably smaller than CMAPs, albeit equally beautiful.

When inserted into a muscle, EMG needle electrodes record whether the muscle fibers are "irritable" or spontaneously active when at rest, and whether motor units are properly activated during a contraction. When resting muscle fibers are abnormal, they develop unexpected spontaneous discharges. These can occur from partial but direct damage to the muscle fibers or from loss of their nerve supply. Moreover, the discharges have very characteristic patterns that electrophysiologists recognize during the recordings. Two of the most common types of discharges are called fibrillations or positive sharp waves and they are most often associated with denervation, loss of the nerve supply to the muscle fiber. Both require EMG to be detected. Electromyographers can observe the characteristic discharges on a screen but they also like to listen to their distinctive sound. Fibrillations sound like raindrops on a tin roof! The intensity of the discharges directly reflects the severity of the denervation or damage. There are other remarkable discharges, such as myotonic discharges, dramatic in persons with myotonic muscular dystrophy. Myotonic discharges make sounds that are thought to sound like a "divebomber"! Myokymic discharges are rhythmic movements of the muscle that are common in the facial muscles of persons with early Guillain Barre syndrome, an acute polyneuropathy we will discuss in more detail. These have a regular "click-click" sound like a record player used to make when it reached the end of the LP side. More types are characterized but their detailed discussion, despite my enthusiasm, is beyond the scope of this brief chapter.

Motor units, or MUPs are analyzed by having the person contract their muscle. The size and number of the MUPs can be assessed. If a person has paralysis from a brain or spinal cord lesion, MUPs may appear normal but the person has difficulty instructing the muscle to generate them, a problem we described earlier called an upper motor neuron lesion. In contrast a person with a nerve or muscle disorder may have very abnormal MUPs. They may be enlarged from collateral sprouting in a partial nerve lesion (see the discussion about this in Chapter 2) whereas they may be small from loss of muscle fibers in a myopathy. In either condition, MUPs may simply not be activated because of the severity of nerve or muscle damage. Recognizing fibrillations, other discharges and abnormal MUPs requires experience.

## Imaging portions of nerves

Nerves are challenging to image not only because of their small size but also because they blend in with the tissues around them. To assemble a complete picture of the nervous system you would need to include every branch of every nerve in muscle, skin, bones and other organs. Ultrasound is currently a favorite on site tool for visualizing important portions of nerves. It is now possible, for example, to use ultrasound to visualize the median nerve at the wrist in carpal tunnel syndrome. Ultrasound has changed some assumptions about nerves in interesting ways. For example, ultrasound images indicate that compressed nerves enlarge and engorge with increases in their blood flow, contrary to what one might assume. You might conclude that a compressed nerve is constricted and suffers ischemia from a compromised blood supply. The clue to making sense of the contradiction is in understanding that compressions are transient, perhaps from repeated bending of the median nerve at the wrist. Once unbent, the nerve responds with swelling and engorgement. Ultrasounds can detect abnormal nerves at other areas of compression and relay whether the nerve is swollen or overgrown. Overgrown "hypertrophic" nerves develop in some types of inherited nerve disorders and in other neuropathies. Primary tumors can rarely arise from nerve trunks. Finally ultrasound can detect if a nerve is physically separated as a result of a neurotmesis injury after trauma. The current advantage of ultrasound is that it is a "point of care" technology, available within the same setting as electrophysiological studies are completed.

MRIs are also used study parts of nerves but the images obtained from this method are currently inadequate. While individual fascicles are sometimes captured, abnormal scans may also simply identify a change in signal within the nerve. Its difficult to discern if this kind of signal change represents inflammation, simple swelling, enlarged nerves or axonal degeneration! Much more work and higher resolution are needed. As above, only large nerve trunks, not distal branches can be carefully studied. MRIs can be very helpful in identifying nerve tumors and cysts or compression from discs or other structures.

## Chop it out: Biopsy of a nerve

You can debate whether removing a nerve to find out what is wrong is a good strategy. This is an older approach, finally standardized in the 1950s–1960s by Drs. Dyck, Asbury, Thomas and others. The importance of standardizing such sampling cannot be overstated. As recently as 10 years ago, it was not uncommon for nerve biopsies to be carried out liberally and sometimes with less care. The quality of some of these samples was substandard and unhelpful for the physician and, more importantly, for the patient. For obvious reasons, removing a nerve is not without penalty. Nerves reside within the body for a purpose and it is not to provide a biopsy. Removing a sensory nerve renders a patch of skin lacking sensation whereas resecting a motor nerve leads to a paralyzed muscle. Over time, investigators realized that only a small number of nerves were easily approached and after biopsy left only limited and tolerable numbness. Moreover, techniques and methodologies were established to conduct biopsies in an optimal manner. The sural nerve behind the ankle is commonly biopsied. If removed from most persons, there would be loss of sensation below the ankle and along the outside of the foot. We mapped out exact territories of sensory loss and tracked how it receded over time in patients having undergone sural nerve biopsy.[41] These areas of loss improved from the outside in, an interesting pattern. Gradual improvement in the area of skin numbness takes place slowly, related to what is called "collateral sprouting," biology that we will discuss further in the section on Regeneration (Chapter 8) below. Sometimes the wounds from the biopsy are painful and slow to heal, especially in persons with diabetes or other conditions. The sural nerve is populated by sensory and some autonomic axons. Studies of the sural nerve yield no insight into motor nerve disorders. Some disorders are also patchy without involvement of the sural nerve. In these instances, sural nerve biopsies are not informative. Rarely other nerves can be biopsied such as the superficial and deep peroneal nerves just above the ankle or the radial sensory nerve at the base of the wrist. The deep peroneal motor nerve contains motor axons supplying the small muscle on top of the foot, the extensor digitorum brevis (the EDB) that extends (lifts up) toes. In the uncommon instance that the deep peroneal nerve is biopsied, the EDB is denervated but fortunately no serious consequence ensues. In many older persons, the EDB is already atrophic, withered and difficult to identify. The deep peroneal nerve is more challenging to identify by the surgeon. Occasionally attempts at nerve biopsy, even of the sural nerve, are a failure, with veins mistaken as nerves.

Despite these critical comments, properly sampled, prepared and analyzed nerve biopsies can provide dramatic and extensive diagnostic information. A careful neurological examination and nerve conduction studies are essential guideposts. Nerves are appropriate to biopsy only if they are involved by the disorder, if they are likely to identify pathology and if treatment is available. For example, if a biopsy identifies vasculitis or amyloidosis, a treatment intervention is anticipated. In these disorders, the biopsy is essential in making a diagnosis and in justifying complex ongoing therapy, sometimes with serious side effects. After the biopsy is taken, the next step requires high quality processing and imaging. Dr. Peter Dyck and colleagues at Mayo clinic have provided essential guidance for routine processing of biopsies so as to extract exquisite anatomical detail (also summarized in Ref. 42). Critical steps include delicate handling to prevent artifacts (handling related changes in the biopsy), the avoidance of dessication (overdrying the nerve), and use of iso-osmolar fixative. Regrettably, many laboratories persist in handling the fixation of nerves incorrectly, as they might handle brain or other tissues; it is placed in hyperosmolar fixative solutions with overly concentrated reagents that distort its structure.[43] Peter Dyck instructed nerve pathologists not only on the unique handling of nerves, but also how to classify their findings.[11] Are the fibers undergoing demyelination, axon damage or both? Examples of sural nerve biopsy images are in Fig. 25.

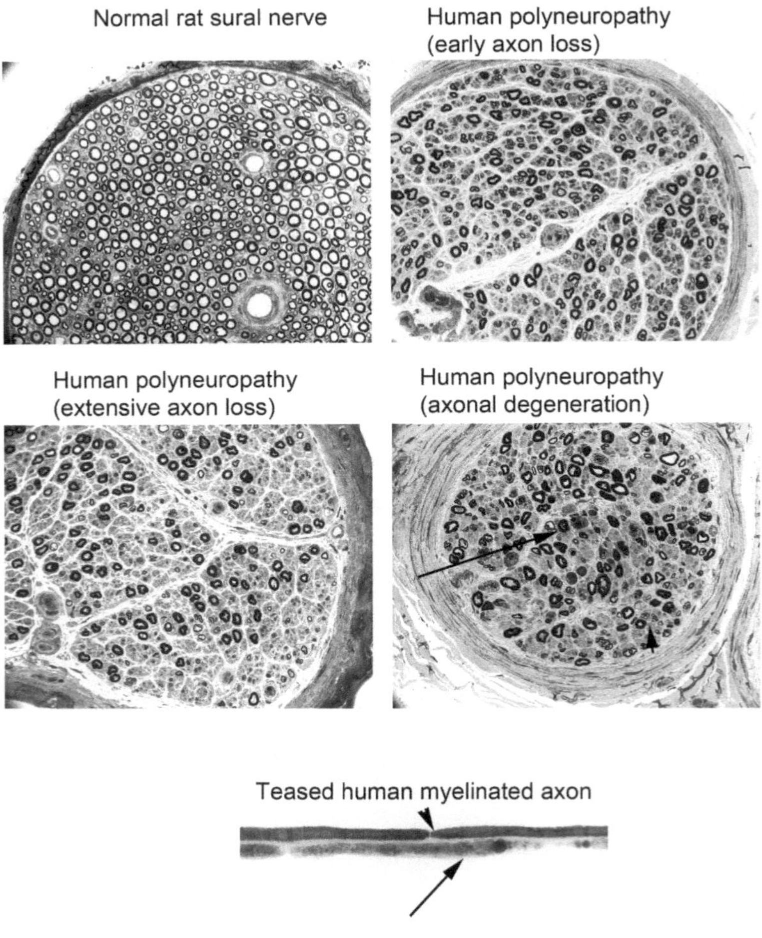


**Fig. 25** Images of sural peripheral nerves in transverse section taken at biopsy. These sections undergo a standardized handling method and are cut in 1 micron sections before staining. Myelinated axons appear as small dark circles. The human samples show degrees of axon loss and axon degeneration (arrow). The sample at the bottom is a teased sample, taken from individual axons laid out longitudinally. In this sample there is a normal myelinated axon with a Node of Ranvier (arrowhead) together with a myelinated axon undergoing early axon degeneration (arrow). The images are taken at 40X, embedded in epon and stained with toluidine blue.

## Detective work: What blood can tell us about nerves

It may not seem plausible that blood tests can tell us a great deal about nerves. However, blood tests can identify whether you are diabetic, have thyroid disease, HIV or an abnormal genetic makeup that all cause neuropathy. They report whether your ethanol (alcohol) level is high, since chronic ethanol use causes neuropathy. If high, you should not drive home from the clinic! Some blood tests, like the CRP (C reactive protein) are elevated in autoimmune conditions that also cause neuropathy. CRP is a nonspecific marker of inflammation and might be elevated in vasculitis, a condition we will hear more about. A blood test for an enzyme called creatine kinase (CK for short) is useful. CK levels rise when muscle tissue is diseased or damaged. In muscular dystrophy, CK levels are very high and in a range of 20000 U/L (units/litre) or higher; the normal level is less than 200 U/L. When larger muscles have lost their nerve supply they also degenerate and CK levels can rise modestly, in the range of 1000–2000 U/L. A new test under development measures the levels of neurofilament in blood or cerebrospinal fluid. Recall that neurofilaments are part of the fine structure of the axon. Its measurement might identify axons that are being rapidly damaged.

Blood tests can detect autoantibodies that are inappropriately directed to our own nerve or SC proteins. There is a growing range of these tests available, often requiring specialized laboratory services. Examples include autoantibodies identified in generalized disorders that also cause neuropathy such as rheumatoid arthritis, systemic lupus erythematosus or celiac disease. There are also new antibodies being discovered that are specific to the peripheral nervous system. Anti-GQ1b antibodies are directed against nerve molecules known as gangliosides. These molecules are composed of lipids and carbohydrate molecules, specifically a glycosphingolipid and a sialic acid, and are found on neural cell surfaces. Anti-GQ1B antibodies are identified in patients with an uncommon inflammation of the peripheral nerves named Miller- Fisher syndrome discussed later. Other autoantibody related disorders have been discovered more recently including anti-Neurofascin 155 and anti-Contactin, both proteins found at Nodes of Ranvier. We will hear more about these problematic "nodopathies." As additional autoantibodies are discovered, we learn more about the causes of other rare neuropathies, and how they might be treated.

Norepinephrine is released from the ends of sympathetic autonomic nerves and enters the blood stream. Its levels can then be measured in the blood. If someone is in the midst of "fright or flite" its levels may be elevated. It might also rise if you feel the need to reprimand a difficult pet.

Of course, high norepinephrine levels usually accompany higher blood pressure, a risk factor for cerebral hemorrhage or other vascular emergencies. In this instance, its better to forgive your pet.

Finally, why would a genetic test using blood tell us anything about nerves? The same DNA is shared by all cells in the body that have nuclei, where DNA resides. Thus, if we want to know our genetic makeup, white blood cells from a blood test will help. Red blood cells do not, because when they are mature they have no nucleus! Given this, why is it that a genetic abnormality does not target all cell types and organs in the body? Why are only nerves damaged, for instance in Charcot-Marie-Tooth (CMT) disease (see below)? This selective targeting is one of the most fascinating aspects of biology. Genes within nuclei await their tasks, but may not be called on to form mRNA. They require "promotion," not unlike a sales event at a new store. Every cell type has a different repertoire of proteins, including those responsible for promotion. Each tissue decides which genes to activate and which not to, whether it's a normal or a mutant gene. In the case of our person with CMT, the SCs usually make a protein called PMP22 that stabilizes myelin sheaths. If PMP22 is missing or mutated in a genetic condition, a demyelinating neuropathy with prominent damage to its myelin results. Fortunately persons with CMT think and act normally, have a normal lifespan and do not usually develop other kinds of disorders. PMP22 is exclusively made in SCs of the PNS because they possess the required promoter proteins. Overall, genetic testing allows us to probe for many genes that can cause neuropathies. Some molecular biology laboratories use specific panels that test for individual changes in DNA associated with neuropathy. More recently clinicians have utilized whole exome sequencing which tests all protein coding portions of DNA for abnormalities, or whole genome sequencing that tests all DNA for changes. The major challenge after performing these tests is interpreting whether the variations detected are meaningful or an unrelated change. This is not the sort of information provided by home kits to test your ancestry or other items; instead it may require input from a geneticist and testing of other family members. Buyer beware!

# CHAPTER 5

# What are neuropathies?

The term "neuropathy" means nerve damage or injury. It has supplanted "neuritis," an older and restricted usage for inflammation of nerve. However we now know that many forms of nerve damage may not involve inflammation. Neuropathy is a more inclusive term that includes direct injury in addition to damage caused by diseases ranging from inherited alterations in nerves to those damaged by an autoimmune attack, to those that follow chemotherapy for cancer. Some neuropathies are named after the human nerve damaged, for example ulnar neuropathy in the arm or peroneal (?fibular for our anatomists) neuropathy in the leg. The term "polyneuropathy" simply means nerve damage that occurs widely in the peripheral nervous system, not localized to single nerve trunks. All axons are susceptible and it is a symmetrical problem, involving both sides of the body. Some may involve mainly motor axons (motor polyneuropathy) or sensory axons (sensory polyneuropathy). Disorders targeting the autonomic nervous system are called autonomic neuropathies. If you immerse yourself into this field, the range and nuances of the wide range of polyneuropathies are engrossing. Some are addressed below.

What if only a single nerve is damaged? Single nerve injuries can be called mononeuropathies, meaning damage to one major nerve or "focal" neuropathy involving of one part of the nerve. Thus, most nerve injuries are "focal" mononeuropathies, but it's a mouthful![44] Not all mononeuropathies are from traumatic injury. They may become compressed by a tumor or cyst, an area of bleeding or swelling from an infection. Chronic damage from a ligament or spinal disc can compress a nerve root in the neck or back. On occasion, nerves may be damaged by interrupting their blood supply. Recall that nerves have a multiplicity of blood vessels, in excess of what they require. Vasculitis for example, is an inflammation of blood vessels, that targets nerve blood vessels and causes ischemic damage to nerves. This is similar to a "stroke," but in this instance the mononeuropathy involves a nerve trunk that lacks adequate blood flow. Infection can also cause mononeuropathies. Leprosy, discussed below, is an infection of peripheral nerves from a bacterium called Mycobacterium leprae. Mononeuropathies are characteristic of

*Our Wired Nerves*
https://doi.org/10.1016/B978-0-12-821487-9.00005-2

leprosy and because the mycobacteria thrive in cooler body areas, superficial skin nerves are colonized. The ears or nose may be sites of infection. If you live in Edmonton you know quite well that these are among the coldest areas of the body and when temperatures plummet to 30 °C below zero they can be frostbitten. However leprosy occurs in the tropics and Edmontonians can rest assured because these temperatures are too cold for even Mycobacterium leprae.

There is one more important term to describe. Mononeuritis multiplex refers to many mononeuropathies. How does this differ from a polyneuropathy? The distinction is that mononeuritis (or mononeuropathy) multiplex (MM) targets individual nerves randomly. Polyneuropathy by definition involves all nerves concurrently. MM is a series of individual hits, not the diffuse problem of a polyneuropathy. Thus a person with mononeuritis multiplex might develop an ulnar neuropathy in the right hand and in the next week a peroneal neuropathy in the left leg, followed in three days by a median neuropathy in the left hand. Other nerves remain completely normal. Why is this important? Only certain disorders exhibit this asymmetric pattern. They can be devastating as in vasculitis, an autoimmune inflammation of small blood vessels mentioned above. Moreover, the condition can be fulminant. This means it can march forward over days to a few weeks rendering irreversible axon damage to several large and important nerves in the body. Early identification and treatment by neurologists are essential. Leprosy and diabetes can also develop chronic MM. Of course if MM is very severe and widespread, it might target most nerves in the body and resemble a severe axonal polyneuropathy. I have followed individuals with MM so severe that they became essentially quadriplegic-all four limbs were paralyzed. We published their story.[45] Aggressive and fulminant MM can result in a very problematic outcome but it is preventable if identified early.

## Mononeuropathies in humans: Intense pain and disability

Mononeuropathies are common, of varied type and disruptive to those who experience them. Despite the fact that damage is strictly confined to one part of the nerve, the pain syndromes associated with mononeuropathies are among the most severe experienced by people. For example, carpal tunnel syndrome, damage to the median nerve at the wrist, can cause incapacitating pain provoked by work. The most common causes of inability to work include carpal tunnel syndrome, and pain from "sciatica," a mononeuropathy that damages a nerve root in the lower spine. While labor associated with birth of a child is often described as the most intense pain a person can experience, the pain from some mononeuropathies may be more intense. This is not my judgment but that of mothers who have experienced both.

Trigeminal neuralgia (TN) is probably the most dramatic example of mononeuropathy pain. "Neuralgia" simply means pain from nerve damage. The trigeminal or fifth cranial nerve is a mixed nerve with sensory and motor components. It exits from the brain and enters the face to supply sensation through its three divisions-the ophthalmic involving forehead and eyelid, the maxillary for the cheek and nose and the mandibular for the jaw and neck. TN is associated with intense shooting pain, most often in the maxillary or ophthalmic divisions of the nerve (Fig. 26). The pain is

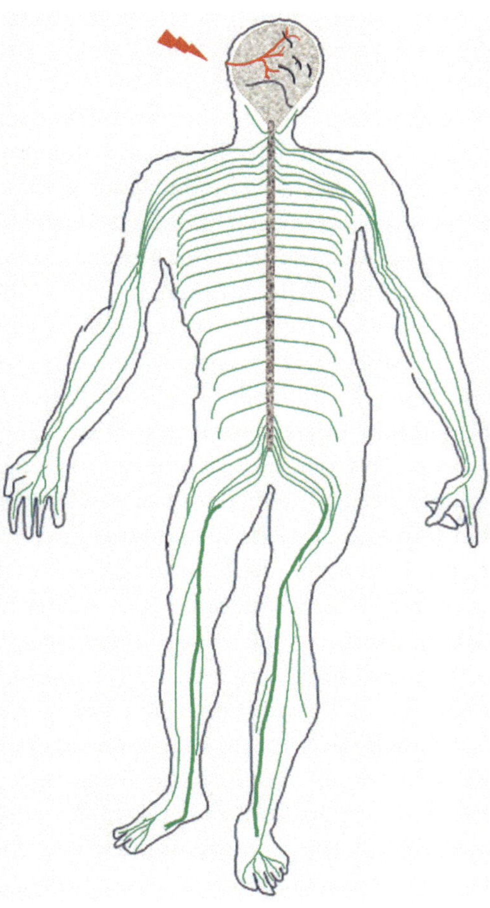

**Fig. 26** Pain localized to the trigeminal nerve in trigeminal neuralgia.

unbearable and repetitive. Moreover it comes in epochs, lasting days to weeks then inexplicably stopping. Intense jabs and waves of pain may be triggered by touching the face, cool air across the face, chewing or laughing. Persons with TN are miserable and standard pain therapies are rarely helpful. Some sufferers have become suicidal and others have taken large doses of narcotics in attempts to relieve pain. However narcotics, specifically opioids like oxycodone, morphine, hydromorphone and others offer only limited benefits but generate serious side effects-confusion, somnolence, constipation or addiction. It is now recognized that many instances of TN arise from compression or irritation of a branch of the nerve in the base of the skull by a nearby extra loop of a blood vessel. This unique association was identified by the late Dr. Peter Jannetta (1932–2016) a Neurosurgeon in Pennsylvania. Despite much early skepticism, decompression of the trigeminal nerve from a vascular loop offers a high rate of cure from TN. The Jannetta procedure is now in use worldwide. Other procedures have been applied to TN including some that deliberately damage the nerve. They have usually been less successful because they replace TN pain with neuropathic pain from the surgical nerve injury! A known complication of these older destructive treatments is "anesthesia dolorosa." This refers to pain despite loss of sensation (anesthesia); despite damage to the nerve, it nonetheless generates intense pain. This curious but difficult outcome will be discussed in more detail below.

Zoster radiculitis or radiculopathy is next on the list of intense and severe mononeuropathies. Zoster is the medical term for "shingles." "Radicular" simply means root, and here refers to a nerve root traveling from the spinal cord in the neck or low back into a peripheral nerve. The herpes zoster virus has a unique and strange predilection to reside in sensory neurons. Originally acquired through exposure to the virus during childhood varicella (chickenpox), it is believed to deposit itself in sensory DRG ganglia of children where it can remain dormant for decades. A later trigger signals the virus to become active, damaging the sensory neuron and spreading down its branches from the nerve root to its branches in the skin. Activation of zoster may occur because of a loss of immune protection from the herpes virus. This may explain why it erupts in older persons with lower levels of immunity or those with immune compromise from cancer, drugs or other causes. The activated virus erupts from the ends of the nerve to form small painful vesicles in the skin in the territory of the infected nerve. Clinicians can identify the targeted nerve simply by observing the distribution of the zoster rash. Most zoster or "shingles" rashes appear wrapped around the trunk (thoracic intercostal nerves) but occasionally they involve limb nerves or nerve roots. The vesicles eventually heal and stop shedding virus, but painful red scars may persist indicating where the inflammation had occurred. Although zoster is usually thought of as a mononeuropathy, several adjacent skin nerves, and their parent ganglia are often involved.

In both TN and zoster radiculopathy sensation in the involved nerve remains curiously intact! Thus despite intense shooting pain from TN or unrelieved burning from zoster, a careful neurological examination identifies normal sensation to light touch and pinprick in the territory of the painful nerve. This paradox highlights the concept that pain may be generated from abnormal nerves with early compression or inflammation. Pain may not require the disruption of axons; they are intact and capable of transmitting information. The principle is that subtle molecular disorganization of a nerve, without frank damage, can generate intense inappropriate forms of pain. The other fascinating part of the story is that despite a small and localized site of nerve damage, the entire nerve tree may alter its properties. Neuron cell bodies in the ganglia, roots and long segments of the nerve are recruited into new and abnormal behavior. The explanation for this unfortunate sequence is unknown.

Carpal tunnel syndrome (CTS) is very common, more frequent in women and impacts both leisure and work (Fig. 27). CTS comes about from repeated nerve small injuries due to wrist flexion and extension. The carpal tunnel is a narrow canal housing the median nerve, bordered by bones and roofed by a firm carpal ligament. Past thinking of this condition, including chronically impaired blood supply, congestion of the veins and others, have been reconsidered. Current advances show that nerves are resilient to changes in their blood vessels and that extensive blood vessel blockade is required to damage a nerve, as might occur in a compartment syndrome or vasculitis. Instead, nerves damaged by CTS are first demyelinated, likely because of telescoping or intussusception of myelin from repeated trauma during wrist movement. Altered myelin architecture triggers SCs and macrophages to remove the sheath and clear the debris. While SCs seek to remake their myelin sheaths, ongoing compression from wrist movement causes further damage. Longstanding CTS also damages axons, likely because AxD is triggered by compression and stretch or crush. Repetitive strain also disrupts the careful partnership between axons and SCs. If myelinated axons lose their relationship with SCs, degeneration may ensue. How exactly CTS interferes with this critical axon-SC relationship warrants further study.

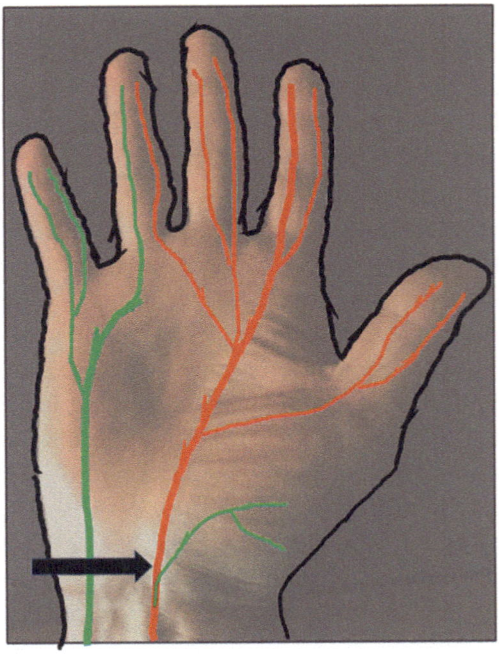

**Fig. 27** Diagram illustrating the site of damage to the median nerve (red) in carpal tunnel syndrome (CTS). Note the sparing of the palmar branch (green) that does not pass through the carpal tunnel. The ulnar nerve is depicted in green.

Needless to say there are not many microscopic studies of the nerves in the carpal tunnel. For obvious reasons a nerve as important as the median nerve can only be removed and studied during an autopsy. The absence of more investigation of this topic is one of those lacunae in our research efforts. Donating your median nerve to science might not be highest on your list if you are creating your will! Despite these unknowns, there are many other unique aspects of CTS that instruct us about nerves, compression, injury and attempts at repair. For example, CTS symptoms seem to be common at night. Does this happen because people sleep on their wrist or place it under a pillow? CTS in the workplace seems to worsen in the evening after a full day of work. It seems unlikely that distraction during the day is enough to alleviate ongoing symptoms. In early and mild CTS there are often symptoms of wrist pain and tingling in the fingers. Demyelinated segments of nerve, or nearby abnormal nodes not only become hyperexcitable but the nerves convince normal neighbors to behave in a similar way. This has been called "ephaptic transmission" meaning that abnormal signals move from one axon to its neighbors. Perhaps this is similar to the stampede that occurs when someone shouts "fire" in a crowded room. Another interesting twist to this story is that many of the first symptoms of CTS do not occur in the fingers supplied by the damaged median nerve. We think of mononeuropathies as causing symptoms confined to the territory of the involved nerve. The median nerve normally supplies sensation to the thumb, index, middle and half of the ring finger. Complete injury to the median nerve causes loss of sensation in this exact territory notwithstanding some minor "normal anatomical variations." CTS patients most often report that their whole hand is numb and that they have to wring or shake their hand to alleviate the symptoms, especially in the mornings. This behavior is almost diagnostic to experienced neurologists. CTS might be diagnosed (almost!) by speaking to someone over the telephone or by seeing the hand wringing on Skype! Thus, although CTS strictly damages median nerve axons, the brain somehow conflates the symptoms into all of the hand. Finally CTS may be surprisingly selective in how it creates symptoms among people and which exact axon branches are damaged. The variations in symptoms are difficult to explain. One possibility is that the exact nerve fascicles that travel to different fingers may be positioned somewhat differently in the wrist among individuals. When compression occurs, the axons destined to different fingers may not sustain identical damage. For example, one person may have numbness of the middle finger, another of the index and another no symptoms at all. Also confusing is that if there are no symptoms, CTS can be identified during nerve conduction studies performed for other diagnostic purposes. Finally, early symptoms often present at the tips of the fingers. Why should this be? Is this an interpretation of abnormal signal by the brain, or do the demyelinated axons at the wrist signal the tips to abnormally fire, much further down the nerve from where the actual damage is? Or, are "tip axons" the only part of the nerve that have undergone a molecular change?

Like so many disorders of nerves, CTS as a problem thought to be well understood. However a deep dive into understanding how it develops, how it progresses and what it does to nerve excitability challenges our thinking. As emphasized in several areas of this book, the apparently "simple" and well understood clinical syndromes are not that simple at all. Several could in themselves generate a lifetime of work to improve our understanding of them. One important message is that knowledge of common disorders like CTS will contribute to our understanding of others including mononeuropathies, and polyneuropathies.

Ulnar neuropathies are common and can impair hand function. The ulnar nerve supplies the skin of the small finger, the inside half of the ring finger, and half of the palm, in addition to many of the hand interosseous (between the bones) muscles. These important muscles help to control finger movements; ulnar neuropathies can be devastating to piano players. The most common cause is pressure over the nerve as it travels across the inside of the elbow into the forearm through what is called the cubital tunnel (Fig. 28). Long distance truck drivers, resting their elbows on the door and computer gamers who rest their elbows on the computer table are at risk!

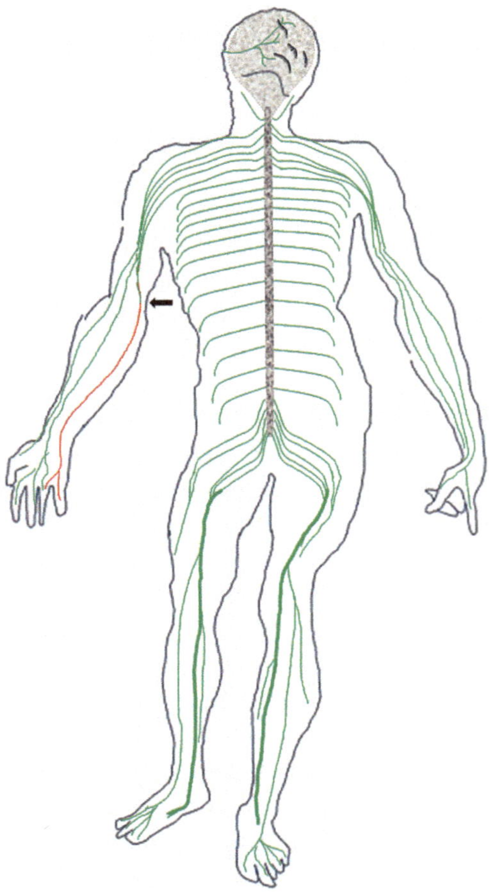

**Fig. 28** Localized area of damage of the ulnar nerve from compression in the cubital tunnel at the elbow.

Radial nerve mononeuropathies are sometimes mistaken for strokes. Recall that a stroke can cause paralysis because an area of the brain is ischemic, deprived of its blood supply. The radial motor axons supply the back of the forearm and extend the wrist and fingers. When injured, the arm develops a wrist drop, an inability to lift up the hand and fingers at the wrist. "Saturday night palsy" is one of the more common types of radial neuropathies. In this case, the nerve is compressed and injured on the inside of the arm above the elbow and intoxication is often involved. Of course the author has also seen this occur after Friday and other nights! The arm is left stretched out over a chair or someone has nestled their head into it. This awkward and poorly chosen position is prolonged over several hours until pain increases and ethanol levels decline! The lesion is usually demyelinating, as in neurapraxia, and can recover after 2–3 months. Since people with this palsy are surprised to awaken with a wrist drop, they are often rushed into an Emergency room. In the ensuing excitement and rush, a brain CT scan may be ordered; the test will be normal because of the absence of any brain lesions. Careful neurological examination tells the story since other muscles in the same arm and hand, not supplied by the radial nerve, are working properly. In contrast, strokes cause paralysis in several muscles because the brain, where strokes occur, codes for movements across multiple nerves and muscles.

Radiculopathy is an equally important mononeuropathy to explore. Typically classified unto themselves, radiculopathies have a history and clinical approach of their own, and are usually treated by spine surgeons and others rather than neurologists. Characterized as damage to nerve roots, they are simply mononeuropathies that occur as axons exit the spine to form nerve trunks. Nerve roots include both motor axons that exit the spinal cord from the ventral root, and sensory axons that exit the dorsal side of the spinal cord. The DRGs are found at the point where the motor and sensory roots join to form the mixed spinal nerve. This area is called the "neural foramina" and its contents are vulnerable to crowding, compression and injury. Vertebrae are complex bony structures and they fit together with the precision of a jigsaw puzzle to form the bony spine. Within their architecture is a central canal for the spinal cord to travel through, and two (for each vertebra) neural formina formed by bony structures above and below it. The foramina, or "passageways" are tunnels that allow the roots to exit and form nerves. The most important and largest nerve roots, and those most prone to damage are in the neck (cervical) [Fig. 29] and low back (lumbar). The vertebrae are separated by flexible discs composed of a tough fibrous capsule that encircles soft, gelatinous disc material.

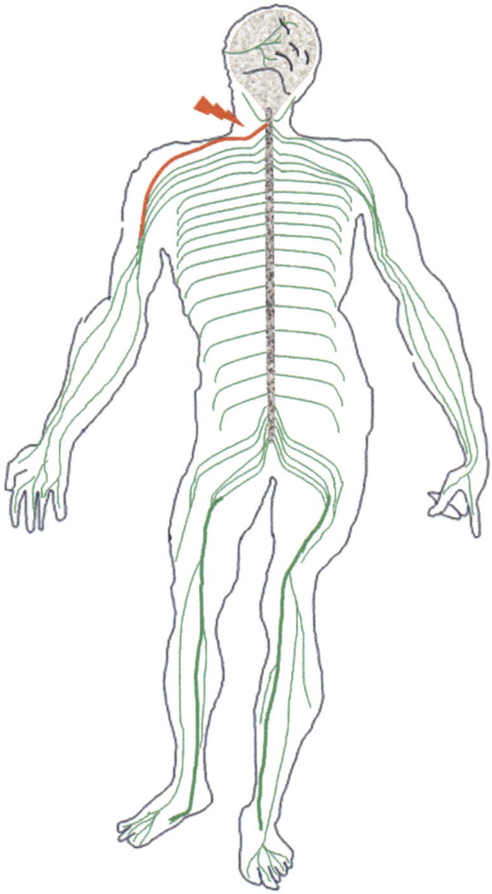

**Fig. 29** Diagram illustrating approximate location of involvement and pain in a C5 cervical radiculopathy, a type of mononeuropathy. The pain is in the neck, radiating into the upper shoulder.

As we age, we are prone to radiculopathies from damage to nerve roots in the small and delicate neural foramina. Bony vertebrae that sustain chronic damage grow unwanted extra material called spurs or "osteophytes" that can compress nerve roots by further crowding available foraminal space. Acute injuries to the back, from heavy lifting, or twisting generate pressure on intervertebral discs. Their tough fibrous capsule of the disc may weaken and bulge or rupture their contents, also called "disc extrusion," impinging the nerve root. Ruptured disc capsules also inflame the nerve roots, perhaps a reason why some anti-inflammatory medications can help symptoms of early "sciatica," a

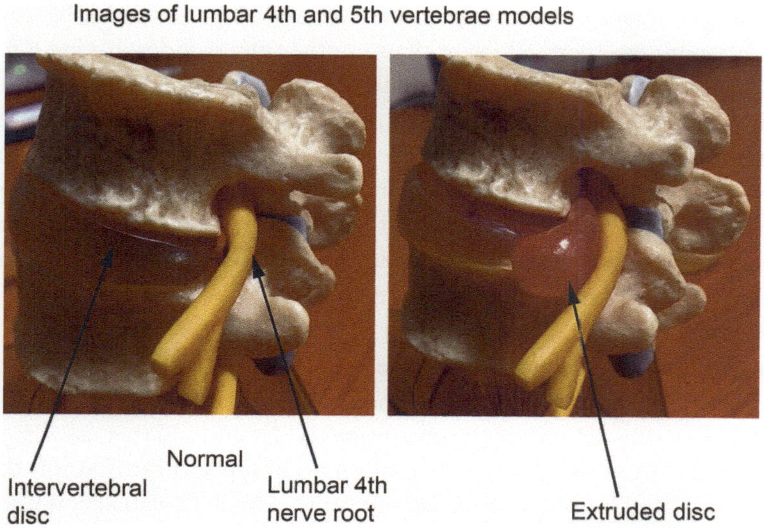


**Fig. 30** Images from a spine model illustrating the passage of the 4th Lumbar nerve root through the intervertebral foramen. The image on the left is normal anatomy. The image on the right shows a large extruded L45 disc impinging on the foramen and nerve root (in this case the lateral disc protrusion compresses the 4th lumbar root). *(Image taken with kind permission of a Kilgore International Corporation human model; Kilgore International, Coldwater MI.)*

termed used to describe the pain from lumbar radiculopathy [Fig. 30]. If you have been sitting reading this book for hours on end, its likely been hard on your discs. It may be better to convince your superior to provide a standing desk so you can keep reading! Mild forms of lumbar nerve root compression may cause pain in the back that "shoots" down the leg, perhaps with tingling, commonly reported as sciatica although the sciatic nerve is not targeted! The pain is worsened by bending the back or flexing the leg at the hip and may be relieved by bed rest, a staple of treatment of "sciatica" for centuries. Disc material compressing nerve roots can be surgically removed, usually by a "laminectomy" in which the lamina in the back part of the vertebra is operated on. This circle of bone and spine protects the spinal cord, but must be opened for the surgeon to identify and remove disc material. New procedures for disc removal, called microdiscectomies involve minimal surgery. Some procedures also involve replacing the damaged disc with artificial disc material placed between vertebrae. Of course, only a small proportion of persons with "sciatica" require surgery since bulging or extruded discs often dry and shrink over time. Symptoms may diminish or disappear provided the person has not worsened the injury-hence bedrest! Back surgery is common and controversial! In Canada, spine surgeons are more reluctant than in other jurisdictions; having a reluctant surgeon only operate when absolutely necessary and a wait list to allow a careful decision is a good thing. Of course if the pain is severe and unremittent or if there is progressive paralysis, surgery may be needed expeditiously. "Failed backs" are described with persons considered for a second, third or fourth lumbar spinal procedure. Unfortunately, with each additional procedure the prognosis for benefit declines. Repeated surgery may augment inflammation and nerve root irritation, worsening the problem instead of helping it.

When a specific root is irritated or damaged, some symptoms match the exact territory of the nerve. Sometimes however, the clinical features can be surprisingly difficult to localize. A classic example involves bulging of the disc between the 4th and 5th lumbar vertebra (L4 and L5) that usually compresses the fifth lumbar nerve root (the L4 compression in Fig. 30 differs because the protrusion is more lateral). With sufficient damage, numbness over the top of the foot and side of the leg below the knee is expected. The person may develop a foot drop from the L5 nerve root damage. In contrast, standing on the toes, or downward (plantar) flexion at the ankle is intact because it is supplied by the first sacral nerve root, S1, one level lower. In contrast "sciatica" pain may seem the same regardless of the exact nerve root that is damaged. While the pattern of the pain can be a guide to which nerve is compressed, it may not be offer a precise localization. Some large disc protrusions can also take out several nerves.

What other mononeuropathies are there and are they important? Virtually every nerve in humans can be targeted and several types of mononeuropathies can involve a given nerve. Some nerves are more vulnerable as they course through limbs and traverse sites prone to cause damage. Bell's palsy is a mononeuropathy of the facial (VIIth) cranial nerve that supplies movement to one side of the face. Damage to the facial nerve causes a facial droop and if severe, a person may not be able to close their eye. The unprotected eye in Bell's palsy can become dry and damaged. Food and liquid might drool out of one side of the mouth because of weakness around the lips. While its also embarrassing to have a facial droop, Bell's palsy is also unique and interesting. It occurs unexpectedly, or "out of the blue" from inflammation of the nerve in a canal, the stylomastoid foramen, through which the nerve travels along the base of the skull from brain to face. Although viral infections such as herpes zoster may trigger it, the inflammation appears to be autoimmune rather than infective. Fortunately Bell's palsy is most often a demyelinating or neurapraxic type of injury and recovery can follow in a few weeks to months. Some persons however develop axonotmesis because of severe and sometimes complete axon damage. These lesions do not recover well. For recovery to occur after axon disruption and AxD, parent axons must send out new branches toward their original targets. This is a slow, often unsuccessful venture and misdirection is among the difficulties illustrated by axonal damage in Bell's palsy. Regenerating motor axons have no clear roadmaps to reconnect to individual facial muscles. Normally, small branches of the facial nerve each connect to small facial muscles, such as the orbicularis oculus that closes the eyes, or the levator labii superioris that raises the lip and several others. Severe Bell's palsy inflames and may effectively transect the axons that form these smaller branches such that regrowth is required. Unfortunately motor axons regrowing beyond the stylomastoid foramen completely lack guidance cues to identify their individual destinations back to the muscle they originally supplied. Hence motor axons make random choices. The consequence is called "aberrant reinnervation" in which regrowing motor axons connect to incorrect muscles. Moreover, the CNS is not aware of this problem and assumes each branch has correctly rejoined its muscle. A person who has had previous axonal damage from Bell's palsy will, over time, misactivate their aberrantly reinnervated facial muscles. If the person is asked to blink, for example, the lip or mouth may instead move. If asked to move the mouth, the person may blink. Another consequence is that the muscles with longstanding loss of their nerve supply become contracted and scarred. Even without attempting to move the face, the person's eyelids on that side of the face may seem narrow. The platysma, a small neck muscle supplied by the facial nerve stands out normally when you grimace or frown (try it!). Some persons having sustained Bell's palsy earlier in their lives may develop a permanent contracture of the platysma and a permanent frown despite their trying to be pleasant!

I would be remiss not to describe how mononeuropathies and the criminal justice system can interface! An interesting but rare mononeuropathy is called "handcuff neuropathy." The cause is vigorous law enforcement, sometimes required, sometimes not. Handcuff neuropathies come in different types, depending on how the alleged criminal is cuffed. It can be from in front with palms facing eachother, behind the back with palms out and perhaps in other creative ways. On one occasion the author met a molecular biologist with ulnar neuropathies of both hands. While his story was not immediately forthcoming, we learned that he had been abusive with law enforcement officers while consuming greater than recommended amounts of ethanol. Not interested in negotiating with this type of customer, the officers cuffed our biologist and admitted him to jail to allow his ethanol to be metabolized. As full mental function returned, he realized that the cuffs felt tight and that he had numbness in his small and ring fingers, areas supplied by the ulnar nerve. A few days later, when evaluated in the neurophysiology laboratory, he was discovered to have local areas of demyelination with conduction block of the ulnar nerves exactly at the cuff site. These were symmetric and very localized neurapraxic injuries! With good behavior, he experienced a full recovery over a period of three months. In this book however, we do not advocate that law enforcement officers routinely use the threat of mononeuropathies to prompt better citizenship.

## What are polyneuropathies: Curious, common and various?

"Polyneuropathy" is not a word likely to raise recognition by most. Contrast this to "stroke," Alzheimer's disease or Parkinson's disease. Many would be surprised to learn that polyneuropathies are as common, and probably more common than any of the above. "Polyneuropathy" simply refers to a disorder that attacks the peripheral nervous system in a general way, beyond involvement of single nerves from a mononeuropathy (Fig. 31). Polyneuropathies do not damage the brain or spinal cord, the central nervous system but the disability rendered is equally severe and often irreversible. Polyneuropathies come in a wide variety of types making their diagnosis and understanding complicated. Imagine a disorder disrupting any one of the many complex parts of the peripheral nerve trunk; each type might be associated with its own type of polyneuropathy. For example, damage of smaller nonmyelinated axons develops in the polyneuropathy complicating HIV infection, sometimes paradoxically worsened by HIV therapy. Severe neuropathic pain is often a feature of "small fiber" polyneuropathies given the normal role of small unmyelinated axons in pain neurotransmission. In contrast, damage to the larger axons present a different set of symptoms and problems. Since larger fibers transmit position sensation and touch pressure to the feet, their involvement causes difficulties in balance and walking. A person with a large fiber polyneuropathy, such as that from chronic alcohol use, may have difficulty walking when the lights are out or when descending stairs. Overall, I am not making a strong case for excessive ethanol use admidst wrist drops from Saturday night palsy and large fiber polyneuropathies. You may elect to forgo its use altogether! Added to this is that ethanol intoxication makes

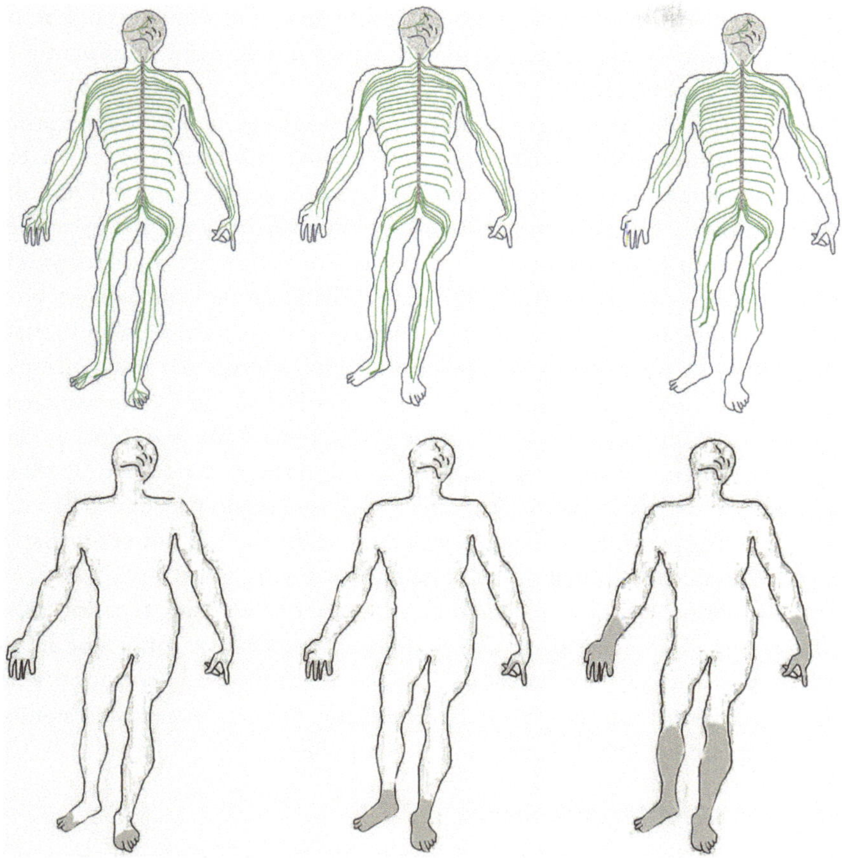

**Fig. 31** Diagrams illustrating progressive loss of axons (upper panels) and sensation (lower panels) in a person with progressive polyneuropathy. Left-initial involvement, Middle-mild early involvement, Right-later more advanced involvement. Note the "stocking and glove" pattern of nerve damage and numbness.

navigating stairs risky without neuropathies. Of course, limited use is not known to cause any mononeuropathies or polyneuropathies, so judge accordingly!

The examples of polyneuropathies discussed above are often called "axonal" because the disorder primarily targets and disrupts axons. Alternatively other types of polyneuropathy target myelin. Since intact myelin is essential for proper transmission of signals, its loss causes conduction block and loss of AP transmission. Some polyneuropathies are chronic and arise from inherited abnormalities in the structure of myelin, whereas others arise from inflammation. An important type of inflammatory demyelinating polyneuropathy, Guillain-Barré syndrome (GBS), arises quickly and fulminantly. It will be discussed below.

With the help of electrophysiological recordings neurologists classify polyneuropathies into axonal and demyelinating subtypes. This is an essential step since most inflammatory demyelinating polyneuropathies, like GBS, fortunately have treatment available. However, many polyneuropathies can also have a complex mixture of both axonal damage and demyelination. These are often the most difficult to diagnose and impacts on how recovery might occur. When a polyneuropathy with weakness, numbness and inability to walk occurs, is it reversible? Finding that the polyneuropathy is largely associated with demyelination but little axon damage suggests eventual reversibility, possibly complete. SCs survive to live another day and to resynthesize myelin sheaths. An exception involves polyneuropathies with inherited deficits in SC function and myelination. While these are also demyelinating polyneuropathies, treatment for them is not yet available. Polyneuropathies with mainly axon damage recovery poorly.

Next follows a discussion touching on some of the major types of polyneuropathies. These are brief and simplified descriptions only but highlight the diversity of their behavior. Some are identified by association with other medical conditions, others have striking presentations and a significant group are inherited genetic polyneuropathies.

## Diabetes mellitus and nerves

Next to global warming, Diabetes Mellitus (DM) is among the most serious threats to humans on our planet. It prevalence is mushrooming. Once listed as having a prevalence of 4–5% of persons, that proportion has doubled in many places such as North America and China[46] (e.g., Canada 7.2%, United States 9.1%, China 9.4%, India 7.8%; 2016 WHO data; [https://www.who.int/diabetes/country-profiles/]). The term "pandemic" has been used to describe the high prevalence and growth of DM, more slowly than an "epidemic" but arguably as serious. It is beyond the scope here to discuss reasons for the explosion of DM worldwide but poor diet, obesity and sedentary lifestyles contribute. Body mass indices, a measure of obesity, are steadily rising in all societies. Its rise is particularly concerning in children.

Type 1 DM is an autoimmune disorder that targets and destroys beta cells in the pancreas. These cells normally produce insulin that is required to control blood glucose levels. This form of DM is severe and life threatening at onset and most type 1 diabetics require insulin therapy for life. However, type 1 DM sufferers are often highly motivated. Some have benefitted from advances in the types of insulin available, insulin delivery pumps and glucose sensing systems or implantation with pancreatic stem cells to cure their disease. One of these approaches, the "Edmonton protocol" was developed by the group at the University of Alberta. The approach involves administering pancreatic stem cells into the portal vein of the abdomen where they are transported into the liver and deposited.[47] Provided that the cells acclimatize in their new home, they begin producing insulin and can potentially reverse the disease.

Type 2 DM is by far the most common form of the disease. At a rate of 85–90% in overall cases of DM, it may appear less severe with initial milder symptoms that are less apparent. Persons with type 2 may have a delay in diagnosis. Disadvantaged populations with poor high fat and carbohydrate diets, less exercise and obesity are prone to develop type 2 DM. Specific groups such as indigenous First Nations persons and Latino families may have a genetic predisposition (Mexico 10.4%). Some parts of the world, such as the Middle East, also have a very high prevalence of type 2 DM (Saudi Arabia 14.4%). In Japan there are high rates of type 2 DM despite less prevalent obesity (10.1% DM, 3.4% obesity compared to 30% obesity in North America; [https://www.who.int/diabetes/country-profiles/]). Despite the availability of treatments, adults with this disease can be difficult to treat. Two important factors that help offset Type 2 DM are weight control and adequate levels of activity. Providing education and strategies for motivation are essential.

Insulin therapy is less effective in lowering elevated glucose levels in persons with type 2 DM, a problem termed "insulin resistance." Insulin synthesis by pancreatic beta cells may be sufficient, with normal or high circulating levels, but fat tissue, muscles and the liver are less sensitive to insulin's actions in taking up glucose. DM is simple to diagnose, with guidelines from the American Diabetes or Canadian Diabetes Associations. For example, two fasting glucose levels of greater than 7.0 mmol/L or an elevated hemoglobin A1C level diagnose DM. Hemoglobin A1C, measured by a blood test, identifies how chronically elevated glucose levels have modified hemoglobin. It indicates whether your glucose levels have been high over several weeks, usually the past 3 months.

Insulin resistance may develop before definite DM is diagnosed and it occurs as part of the "metabolic syndrome." It includes obesity, insulin resistance and high lipid levels and it is associated with higher risks of cardiovascular disease such as myocardial infarction and stroke. Other terms describing "prediabetes" are glucose intolerance, and abnormal glucose metabolism, often diagnosed by a series of glucose measurements carried out after ingesting a glucose load, known as a glucose tolerance test or GTT. Irrespective of the name, complications like polyneuropathy are now known to develop before frank DM is present. However, it may also be an opportunity to intervene. While debated, many believe that the early damage of nerves in metabolic syndrome can recover with weight loss and better nutrition alone.

There are several forms of therapy for type 2 DM, not reviewed here, but insulin is frequently required as the disorder progresses. In longer standing type 2 DM, beta cells eventually fail and no longer produce sufficient insulin. Thus a person with type 2 DM may progress through stages having high insulin levels and insulin resistance, to later severe loss of pancreatic insulin production. Therapeutic programs to lower glucose levels to a normal range are designed to achieve "diabetic control" but many fail despite significant effort. Despite treatment, and improved glucose levels, complications of DM are common, extensive and often irreversible. For example DM is the most common cause of kidney failure requiring dialysis and the most common cause of blindness from retinal damage. Beta stem cell transplants, insulin pumps and other forms of new technology are largely targeted toward type 1 DM and may not be appropriate or helpful in those with type 2 DM. For example, implanting beta stem cells may generate higher levels of insulin in a type 2 DM person, but will fail to reverse insulin resistance.

Diabetic polyneuropathy (DPN) is one of the most common complications of DM. It develops in both types 1 and 2 DM. Perhaps surprisingly, DPN can be detected in over 50% of all persons with DM! DPN does not cause symptoms in everyone with DM but it does in 20–30%, also a high and concerning proportion. Symptoms include neuropathic pain, sometimes very severe, loss of sensation, autonomic abnormalities and later motor weakness. Over the last 50 years, the prevalence of DPN has surpassed all other causes of polyneuropathy, including leprosy, at one time the most common peripheral nerve disorder. Given the worldwide pandemic of type 2 DM, DPN exceeds the combined prevalence of MS, Parkinson's disease and ALS by over a factor of 10! I routinely remind my neurological colleagues that there is something out there, DPN, that dwarfs the condition they focus on. This helps to foster some interprofessional rivalry.

The onset of DPN or other DM complications is variable. Persons with longstanding type 1 DM that was diagnosed decades earlier can be remarkably free of complications. Many have been diligent and achieved very tight glucose control. Others, including some children seem to develop complications early in the courses of DM. While we believe that poor diabetic control accelerates overall complications their appearance can be surprisingly uneven. One person, for example, may have severe retinopathy but normal kidneys and no neuropathy. Another may experience severe neuropathy but no retinopathy or kidney damage (nephropathy). Genetic factors, some that are known and some not yet discovered, predispose people to developing complications and may influence the rapidity of their course. Some investigators have suggested that elevated cholesterol or triglycerides, combined with high glucose levels predispose individuals to earlier DPN. Others suggest it may be fluctuations in glucose levels, such as the spikes in levels after meals that are responsible. Chronically low glucose levels can also cause a less common polyneuropathy, known as hypoglycemic neuropathy! Finally, it may be lack of insulin, or its failure to support neurons directly during DM, that contributes to the development of neuropathy.

In most cases, DPN first targets the longest nerves, a general theme of polyneuropathies. Our longest nerves are those that travel from the spinal canal in the low back to the ends of our feet. In DPN, one of the first symptoms may be pain in the feet. The sensation can be intense, despite the fact that permanent nerve damage is not yet present. Like trigeminal and zoster neuropathies, sensation remains normal but severe pain is experienced. The pain in DPN can be spontaneous, without a trigger or brought on by touch or walking. DPN sufferers may notice pain arising from the ends of their toes or in their soles that travels upward to the forefoot, ankle or knee depending on the severity. Many words are used to describe these symptoms. One of the most striking images that illustrates the pain was used in an industry advertisement a number of years ago. It highlights a DPN patient lying in bed with their feet from the ankles out depicted in red, as if they are on fire, with a cradle to hold the covers off of their feet. Even the most innocuous touch, such as that from bed covers, is not tolerated. "My feet feel as if they are on fire!" is a common description and the pain is often worse at night. In contrast during the day, it may be that nerves are constantly activated by socks, shoes, walking so as to distract from the pain. It could also be that a change in diurnal hormones at night serves to activate inappropriate axon discharges. The overall impact is interrupted sleep, impaired job performance and activities of daily living and depression. Some who suffer from DPN ask whether their feet can be amputated to relieve the pain. This gives you a "sense" of how bad the pain may be and to what ends people seek relief. Unfortunately amputation may offer no respite since the neuropathic pain continues to be generated by remaining nerves targeted by DPN.

Other words are also used to describe the pain from DPN. They include electrical sensations or shocks in the feet, sometime provoked by walking, but at other times "out of the blue." The feet can feel tight as if they have constricting footware. Prickling feelings, pins and needles and "jolts" of pain are common descriptors as is deep aching pain. Why do all of these words apply to the same disorder? This is part of the complexity of medicine, neurology and peripheral neurology. Every individual has a different story. It is also the role of the health care provider to sort out what the symptoms represent. Some of the variation occurs as a result of types of sensory axons that are damaged first. For example, if axons that normally transmit heat or cold are first damaged, burning might be the consequence. If larger axons that transmit light touch are involved, perhaps pins and needles sensations occur. Thus we can begin to understand why individual A may have burning feet while B might have deep aching and C no pain whatsoever. This is despite all three persons having had DM for the same period of time and of same severity. Patient D might have previously had pain, but it eventually disappeared, replaced by numb feet without feeling.

Persons with DPN sometimes have difficulty knowing whether they prefer pain or numbness. Tragically, one of the most common scenarios is a combination. The patient with DPN may have numb feet but continue to experience pain. How is this possible? One might consider that DPN is a progressive problem that involves a sequence of events. First are alterations in the protein structure of the nerve endings. Instead of having a normal mix of receptors to heat and cold for example, this mix may be changed and altered sensory messages are sent. Other proteins in the nerve that normally inhibit these sensing proteins may be missing or damaged and nerve terminals begin to fire pain impulses inappropriately. The normal restraints are missing. An interesting question here is whether nerves firing excessive pain impulses are also predegenerative, on the cusp of AxD and "dying back." In nerve endings that lack energy reserves to maintain an ionic balance, excessive sodium ions and prolonged depolarization may generate inappropriate APs. Abnormalities of nerve mitochondria have been implicated in this scenario but whether their changes are cause or effect is debated. Over time abnormal physiology, for example rises in axon calcium levels, might lead to axon death or retraction. Another possibility is that the protein changes in sensory axons and neurons that cause pain are a separate but unlucky development, different than degeneration.

"Dying back" is an important concept, first described in polyneuropathies from chemical toxic exposure.[48] The idea is that the extended endings of the neuron tree, with axons that stretch out to the feet are not able to sustain themselves by axoplasmic transport from the cell body. Failure of support from parent sensory neurons of the DRG that are compromised by disease results in terminals that "die back" or retract. As the distant axon terminals disappear, the patient develops numbness. However, the entire axon tree is also abnormal and may continue to generate inappropriate pain until it too degenerates over time. One positive note about the "dying back" idea (if there is something positive about it) is that the cell body remains alive and intact. Keeping this in mind, if the polyneuropathy could be arrested, preserved cell bodies might have the capability to form new regenerating axons in order to restore sensation.

Investigators have sampled small biopsies of the skin to examine axon endings in DPN. As predicted, the small axons that grow or "ramify" through the epidermis begin to disappear in persons with DPN. While their loss is greatest in the lower leg or foot, early loss can also be detected at higher levels, for example in the thigh. More severe DPN involves the fingers and hands. This further illustrates the theme that the longest nerves in polyneuropathies are usually targeted first, also described as "length dependent" neurological damage. Residual axons develop swellings known as endbulbs. These may generate abnormal pain discharges, possibly from their accumulation of excess sodium ion channels.[49,50]

Pain then numbness or both are not the only symptoms of DPN. You may recall that the autonomic nervous system also sends axons that supply sweat glands out into the skin. At the same time that sensory nerves "die back," autonomic sweat fibers in the skin may also disappear. Patients may notice that their socks no longer get damp at the end of the day and are dry and odorless! This contrasts with their trunk and back where profuse sweating can occur. Some clinics have specialized laboratories for testing sweating in patients. The "thermoregulatory sweat test" involves wheeling a person into a heated sweat chamber or "cradle" where a sweating patterns can be mapped. Once suitably warmed and sweating, the subject is sprinkled with powder over the skin that changes into a color when exposed to sweat. An early version of this used starch and iodine that turned black when exposed to sweat. Areas of the skin that do not turn color after warming and application of the indicator dye, are unable to sweat (anhidrosis). In a DPN patient, there may be no sweat stain over the feet to the ankles and in the hands, a "stocking and glove" pattern.

Many persons with DM are worried about their risk of amputation. Some interpret the "pins and needles" sensations in their feet as "poor circulation." As discussed however, these symptoms arise because of polyneuropathy and abnormal discharges of sensory nerves. Nonetheless persons with DM are at risk for amputation as a result of damage to small blood vessels and insensitivity of the skin to injury from DPN. Unrecognized injury to insensitive feet leads to skin breakdown, deep ulceration, and bacterial infection. Without amputation, infection can spread and lead to death. Diabetic specialists relay sobering stories of their DM patients walking about with tight boots oblivious to the damage being caused. When the boots are removed with difficulty, severe skin breakdown is discovered. Most health care workers in DM clinics recommend having someone inspect your feet daily. There can be surprises, such as a nail in your sole without your having realizing it!

What of motor nerves in DPN? Like sensory nerves, their endings can also degenerate but more slowly. Why diabetic sensory and perhaps autonomic axons are more vulnerable than motor nerves is not explained. Their differing molecular properties and anatomy may contribute to their apparent resilience. Motor neurons also compensate for the loss of their axons, hiding the extent of damage. For example, we discussed earlier that we can lose a substantial proportion of our motor axons before we notice weakness.[51] Eventual damage to motor nerves can cause wasting of the small muscles of the foot and deformity, changes that make feet yet more vulnerable to pressure damage from walking or standing. Since the feet are not only deformed but also insensitive, injury and ulcers may not be recognized. The small muscles of the foot also provide stability when we walk or stand. Their loss contributes to poor balance and falling. As DPN worsens over time, muscles higher up the leg and in the hands become weak and atrophic resulting in bilateral "foot drop," weakness and wasting.

## More types of diabetic nerve damage

Our discussion so far has been about diabetic polyneuropathy (DPN) a symmetrical condition with feet and hands subject to pain, loss of sensation and sweating and later muscle weakness. Diabetic persons are also susceptible to several mononeuropathies. These include carpal tunnel syndrome, ulnar neuropathy at the elbow and others. These behave in a similar fashion to mononeuropathies in nondiabetic persons but recover more slowly with treatment.

Diabetics also develop several unusual mononeuropathies. Damage to the third cranial nerve, a third cranial neuropathy, causes the eyelid to droop and impairs eye movement. Eyelid opening, pupil size, upward, downward and inward movements of the eye are each under control of the third cranial nerve. A number of disorders, including expanding brain blood vessel aneurysms can damage it. Since the third cranial nerve constricts the pupil, these mononeuropathies usually cause the pupil to dilate, excepting when DM is responsible, a helpful distinction. DM only damages axons within central portions the nerve, sparing the axons for the pupil that are found in the outer layers of the nerve.[52] Another uncommon mononeuropathy in DM involves small "cutaneous" nerves that supply sensation to the skin around the chest or on the abdomen.[53] They arise from the thoracic and upper lumbar spinal cord, exit the spinal canal, wind around the trunk and enter the skin. When damaged in DM, there is excruciating pain from the chest or abdomen. Undiagnosed, this pain might prompt a visit to the Emergency room and use of tests like CT scans. Rarely patients were even operated on because of the concern over a serious chest or abdominal emergency. It is critical to recognize that the problem is actually in the nerves of the skin on the outside rather than inside! The condition resembles zoster or shingles pain, involves similar nerves but does not form skin vesicles containing herpes virus particles.

Plexopathy is also an unusual but severe DM mononeuropathy. Recall that the nerve roots form up into plexi (plexi is plural, plexus, singular), like highway interchanges, before they branch into nerve trunks. Since this mononeuropathy usually targets the lumbosacral plexus, a group of nerves traveling to the leg, it is named diabetic lumbosacral radiculoplexopathy or plexopathy; quite a mouthful! A simpler term is "Bruns Garland" syndrome. This complication can be quite severe and begins with deep aching in one thigh, followed by loss or atrophy of the thigh muscles and eventually weakness. The pain and weakness may be so severe that the patient can no longer walk. Despite the location of the plexus damage, numbness does not occur and the motor weakness also slowly improves over time. Yet more unexpected is that it can first develop after insulin is started or when glucose levels improve. This is a little difficult to understand since most complications of DM relate to poor diabetic control whereas diabetic lumbosacral plexopathy seems to be provoked by significant improvements in glucose levels. Of course, fear of this rare complication should not be considered as an excuse for poor diabetic control. We do not know why damage at this site should develop, seemingly "out of the blue." The rare autopsy studies completed in patients with it have suggested it could be caused by inflammation and damage to the plexus blood supply.[54,55]

Autonomic nerve damage in DM can happen in the early or later stages of the disease. You will recall that autonomic nerves connect to many organs in the body. Autonomic neuropathy, or DM damage to autonomic nerves is wide ranging. We have already discussed loss of sweating. One of the more serious types of autonomic neuropathy is called cardiac autonomic neuropathy because the nerves to the heart retract and eventually disappear. This can be so severe that no nerve supply remains, not unlike a newly transplanted heart. Normally when a donor heart is implanted, its nerve supply is severed upon removal from the donor and in the recipient, the new heart operates without input from autonomic nerves. While the heart is capable of functioning in this way, it cannot adjust the rate of its rhythm without signals from autonomic nerves. In diabetes, varying degrees of autonomic axon loss characterize cardiac autonomic neuropathy. Partial axon loss may be more dangerous than complete cardiac denervation, given that some portions of the heart might receive signals whereas others do not. DM subjects are at risk from serious cardiac rhythm abnormalities including abnormal EKG electrical patterns that predict sudden death! Overall cardiac autonomic neuropathy is thought to be a risk factor for premature death in diabetics.

Another autonomic complication is low blood pressure on standing, called "postural hypotension." With this problem, a diabetic person's blood pressure falls when they stand up. They can feel dizzy, sweaty, have difficulty walking or actually lose consciousness. Postural hypotension develops because the small autonomic nerves that constrict blood vessels in the lower legs and abdomen fail. Usually when we stand, these small blood vessels respond with constriction to keep the blood from pooling in the lower part of the body. The sympathetic autonomic nerves signal the blood vessels to constrict, to maintain blood pressure when we stand. As a result of diabetic autonomic neuropathy, these nerves may be damaged or missing, blood pools on standing, the blood pressure falls and there may be fainting. If the person is yet more unlucky, a fall and injury might ensue.

Autonomic neuropathy can damage the nerve supply of the gastrointestinal system. It impacts how the esophagus contracts when we eat or how the stomach directs food into the duodenum. A classic symptom of diabetic autonomic neuropathy is delayed gastric (stomach) emptying. The stomach feels full and we experience loss of appetite. We might feel bloated, have bad breath or nausea and stomach pain. The neuropathy can also involve the small and large intestines leading to malabsorption of food and diarrhea.

Sometimes there is constipation that can be severe enough to require disimpaction! For those who have not worked in the trenches of health care, this involves using a gloved finger to manually remove stool that cannot be expelled. Carrying out a disimpaction at one time was a right of passage learning procedure for ward house staff and nurses! At other times persons with diabetic gastrointestinal autonomic neuropathy may have diarrhea, worse at night or often alternating constipation and diarrhea. There may be incontinence, loss of control of the bowel.

The urinary system is also impacted by diabetic autonomic neuropathy. The bladder may develop insensitivity to its content of urine. As a result, it enlarges and begins to overflow, "overflow incontinence." More commonly abnormal bladder function causes urgency and fullness in diabetics. It is important to remember, that many other conditions, including certain types of medication use can cause symptoms that resemble diabetic autonomic neuropathy. These reversible causes of autonomic difficulty should not be missed since they can be easily corrected!

Diabetes is the most common cause of erection failure, known as erectile dysfunction (ED) in men. Up to 50% of diabetic men experience this problem. New medications are available but not always therapeutic. ED develops because the autonomic nerves fail to dilate the blood channels in the penis. This is required for sustained engorgement and erection. Women with diabetes can also report difficulties with sexual activity including failure of lubrication, pain and other issues.

There are yet other unusual abnormalities caused by diabetic autonomic neuropathy. The size of the pupils is controlled by sympathetic and parasympathetic autonomic nerves. These dilate or constrict the pupil. When we feel excited our sympathetic nerves are activated and our pupils dilate. In DM, the pupil reactions may be sluggish or sometimes completely absent! If a diabetic is unconscious do not assume they are brain dead!

Another odd symptom of diabetic autonomic neuropathy is "hypoglycemic unawareness." When our glucose levels decline, we feel dizzy, sweaty and generally unwell. The treatment is to eat. In a diabetic, none of this may happen. Yet more concerning is the observation that each time a diabetic person has low glucose levels, the worse this problem becomes. Repeated episodes of hypoglycemic unawareness increases the severity of subsequent episodes. Given this, a diabetic person might be walking about confused and unwell, while completely unaware that their glucose levels are low. A glucose level of less than 2.7 mmol/L (normal should be 5.0 or greater), causes brain malfunction.

Beyond improving glucose control, there are no options at this time to reverse or prevent diabetic neuropathies from developing. Adjustments in insulin dosing, addition of other glucose lowering agents and other strategies that achieve better control reduce the development and progression of DPN. Many specific approaches toward neuropathy treatment have been tried. For example, agents known as "aldose reductase inhibitors" (ARIs) were trialed over three decades in attempts to prevent or reduce the severity of DPN. ARIs prevent nerves from accumulating sugar alcohols known as "polyols" but their overall impact has been marginal.[56] Several ARIs caused serious side effects. Vitamins, antioxidants and other treatments have not offered impressive benefits. Despite this dismal record, therapy for neuropathic pain is available, is of partial benefit and is discussed below. Given all of this, the onus on investigators has been to find treatments that prevent damage and allow the nerves to regrow. There is some success in mice, but not yet in humans.

## Curing diabetic neuropathy in rodents

Diabetic rats, mice and cats tell us something about human DM. To mimic human disease, diabetes can be induced in rats and mice by an injection of stretozotocin (STZ), a chemotherapy drug that damages islet cells. Knowing the precise onset of DM, usually within 5 days of injection, allows careful timing of proposed treatments. There are other forms of genetic DM types 1 and 2 in mice or rats that can be studied. Cats develop spontaneous autoimmune type 1 DM, but since they are important companions, they are not usually studied, unless requested by a cat veterinarian. Like humans, DPN in these animal models develop loss of sensation to heat or touch, loss of epidermal axons and abnormalities of nerve conduction.

What have the rodent studies taught us? A great deal. They have allowed us to examine a range of pathways and approaches for the treatment of DPN. Both their successes and failures have been informative. We have learned that diabetic neurons have high levels of polyols prompting the clinical testing of ARIs described above. A newer story involves AGEs, or "advanced glycosylation endproducts."[57] These are cellular proteins that are permanently altered because of their exposure to glucose over time periods of months to years. Blocking AGEs from forming might prevent DPN. What the models have also demonstrated is how mitochondria, the energy factories of neurons, fail to work properly in DPN. Some investigators believe this might be the entire basis for the axon damage of DPN.[58] An equally plausible line of inquiry investigates insulin, a major area of study in our own laboratory. While we think of insulin as lowering glucose levels, it can also attach to insulin receptors on neurons and create growth signals.[59] Insulin is a very potent neuron growth molecule.[60] Could DPN develop from a lack of insulin to support neurons? Following along this idea is the concept that neurons, to stay alive and healthy, need a constant supply of growth molecules such as that provided by circulating insulin in the bloodstream. Since type 1 DM results from loss of insulin production by beta cells in the pancreas, nerves might suffer the consequences of insulin deficiency. However, in type 2 DM, neurons might not respond to insulin because of "resistance" that we described earlier. Insulin resistance is also now known to develop in neurons that lose their ability to grow in response to insulin[61] and the challenge would be to find how to restore their sensitivity. Experimental work has shown that in type 1 DM models, small doses of insulin, less than required to impact blood glucose levels, can be administered directly to axon insulin receptors. In these models low dose direct insulin applied in this way reversed changes in conduction velocity, restored skin sensation and replenished skin epidermal axons. Overall it may be that maintaining essential growth signals in neurons through insulin could be an alternative strategy to treat DPN in humans.

Another promising line of research has examined how DM changes genes in neurons.[62] Studies including our own suggest that there are profound genetic changes in neurons that encode proteins we previously knew nothing about. We do not know if these approaches will identify a "magic bullet" for curing DPN, but it could be an essential step forward. This is an example of why fundamental basic research, also called discovery research, is critical to moving the field. The health costs of DPN, from pain to amputation probably exceed the cost of new research by a factor of a million to one or more! We estimated that to double the budget for this kind of biomedical discovery research in Canada, would require Canadians to pay $30 extra a year in taxes, far less than the cost of a concert!

## Polyneuropathies with inflammation

Several types of difficult polyneuropathy are caused by inflammation within nerves. Some polyneuropathies or mononeuropathies, like leprosy and shingles (Herpes zoster) develop because a bacterium or virus directly invades the SC or nerve. However, I will also discuss some more common neuropathies with inflammation that do not involve direct infection of nerves. These are instead instances of nerve inflammation because of an autoimmune attack.

Examples are Guillain-Barré syndrome and CIDP (chronic inflammatory demyelinating polyneuropathy).

## *Leprosy*

Leprosy is probably the oldest neuropathy recognized, with considerable history attached to it dating back over a millennium. Leprosy is also called "Hansen's disease" named after Gerhard Amauer Hansen, a Norwegian physician (1841–1912). Perhaps it is best we not pay too close attention to the eponym in this instance. Hansen not only entered into a bitter dispute over who discovered the bacterium with his German colleague Albert Neisser, but was also apparently dismissed for trying to infect a patient with Mycobacterium leprae (https://en.wikipedia.org/wiki/Gerhard_Armauer_Hansen)! Leprosy is a complicated disorder, not only for its biology but also because of its longstanding social impact. At one time it was thought to be the most common neuropathy worldwide, but in reality it is dwarfed in prevalence by neuropathy associated with diabetes. Despite attempts to eradicate leprosy, which by WHO means less than one case per 100,000 persons, it is still a problem. India, Brazil, Indonesia and Nigeria continue to have substantial numbers of persons with leprosy.[63] Multidrug therapy (MDT) has cured the infection but has not restored or regenerated the nerve damage in persons who have sustained leprous neuropathy. Leprosy has been a severe social stigma for centuries, and it is continuing today. Persons with leprosy are exiled from their communities, driven to live separately in colonies, and forced into subsistence living. Through laudable activist programs involving the WHO and philanthropists such as Yohei Sasakawa[64] however, this is changing. Education of governments and people to eliminate the fear and discrimination associated with leprosy is lessening the stigma.

In leprosy there may be intense inflammation within the nerve because the bacterium, Mycobacterium leprae, invades SCs. In fact, these bacteria not only prefer SCs, they are yet more entitled and prefer SCs associated with unmyelinated, not myelinated axons. This may explain why leprosy particularly involves small unmyelinated axons of the skin. Macrophages are also among the favorite hosts of the bacterium. A further interesting factoid is that armadillos may host the disorder, although for humans to acquire leprosy, armadillos would not the primary source! When is the last time you interacted with an armadillo? Mycobacterium leprae is actually difficult to transmit, likely through nasal secretions, possibly close skin contact or blood. The important point is that the disorder is much more difficult to contract than generally feared. It is also not sexually transmitted. The incubation period for leprosy is very long, ranging from 10 to 20 years.[63] Activists that work hard to eliminate the fear of leprosy routinely make a point of shaking hands with those suffering from and stigmatized by leprosy.[64] They do not contract the disease from this simple human gesture and it has often reassured decision makers. It is also a very kind and meaningful action to take for someone who has been shunned from society.

Depending on the immune system of the host, leprosy can act in different ways. In a proportion of persons, the immune system is overwhelmed by Mycobacterium leprae and the nerves, SCs and skin host large numbers of the bacterium. This is known as lepromatous leprosy or multibacillary (many bacteria) leprosy. Despite an inadequate immune defence, the disorder can attack and inflame single nerves or multiple nerves. One of the interesting predilections of Mycobacterium leprae is for cooler parts of the body, preferably 27 to 30 °C.[63] This is why some chronic leprosy patients have severe facial deformities; the bacterium invades and collapses the nasal cartilage. Loss of digits results from their susceptibility to damage due to insensitivity. In other persons, there is an intense immune reaction to invasion of the bacterium into the skin and SCs.[65] In this situation, few bacteria can be detected microscopically but there is considerable nerve and skin damage because of the raging inflammation that ensues. This form is known as tuberculoid or paucibacillary (few bacteria) leprosy. Like many issues (!), the reality is more complex and specialists in leprosy recognize additional subtypes such as borderline lepromatous (more like lepromatous leprosy but with fewer bacteria and more inflammation), borderline leprosy (in between the lepromatous and tuberculoid types) and borderline tuberculoid (more like tuberculoid but more bacteria present).

Leprosy evolves into a chronic ongoing battle between the bacterium and the immune system. There may be swollen inflamed nerves damaged both by immune cells and by the infection. Since the bacterium invades cooler nerves, they can become enlarged and visible beneath the skin. Many forms of leprosy start out as mononeuropathies, such as ulnar neuropathy. "Mononeuritis multiplex," referring to multiple mononeuropathies, most often characterizes advancing leprous neuropathy.[65–67] Diffuse and symmetrical polyneuropathies also develop in leprosy. In general, neuropathy is accompanied by pale patches of skin that are insensitive because they have lost their skin nerve supply.[65] However, some types of leprosy are confined to nerves, not involving skin, and are called the pure "neuritic" form. Yet others are classified as "silent," also called "quiet nerve paralysis," since they spare the skin but also develop without numbness, pain, paralysis or weakness.[68] Why would this be a problem? The issue is that quiet nerve paralysis "catches up" with the patient, eventually rendering all of the disabilities associated with neuropathies, but with much less warning.

Diagnosing leprosy requires an astute clinician, a skin smear demonstrating mycobacteria or occasionally a nerve biopsy. The biopsies, particularly of enlarged swollen nerves, discover changes that are truly alarming, with few normal axons remaining, thickened perineurium and eventual fibrosis ("scarring"). Residual axons may be demyelinated or atrophic (shrunken) and are surrounded by inflammatory cells such as macrophages.[66] Depending on the type of leprosy, it may be possible to detect many, few or no Mycobacterium leprae in the nerve. Finally, despite having MDT, patients may have an acute reaction that worsens their condition during treatment. Some mononeuropathies may require surgical release at sites of scarring, albeit of uncertain benefit. Most persons will require rehabilitation.[66]

### *Guillain-Barré syndrome*

Probably the most dramatic polyneuropathy of all is Guillain-Barré Syndrome (GBS) named after two ancient French neurologists, Georges Guillain (1876–1961) and Jean Alexandre Barré (1880–1967) who wrote about the disorder among soldiers in 1916. However in the spirit of eponym confusion, GBS was actually described earlier, in 1859, by Jean Landry (1826–1865) who was also French. Finally, to make matters worse, Andre Strohl (1887–1977), the junior coauthor on the 1916 paper, is usually omitted from the eponym.

An eponym refers the name of person, in our cases physicians or scientists, associated with a major discovery or finding. There is a dislike for applying eponyms in Medicine these days since they generate disagreement over contribution and are considered a means to inflate already enlarged egos. Many have been discarded, particularly if named after someone whose contributions masked a darker side, such as unethical experiments on people. I will only argue for one new one later in the book.

GBS is an autoimmune attack on peripheral nerves, most often myelin or a ganglioside associated with myelin or axons. It is triggered by a viral or bacterial infection that typically occurs two weeks earlier before GBS develops. Virus invasion of the nerve does not occur. Why should a common infection occasionally trigger GBS in someone? The explanation for this form of self rejection involves an immune system that is inappropriately redirected to attack self instead of a virus or bacterium. "Molecular mimicry" describes the concept that certain three dimensionally folded viral or bacteria proteins have an uncanny resemblance to normal nerve structure. Through this resemblance, the immune system is reprogrammed to attack myelin, axons or gangliosides. A preceding infection thereby primes the immune system to generate such an attack on self instead of an invading organism. Treatments that suppress this unwanted inflammation help to speed recovery from GBS.

GBS is a striking problem. Persons with GBS may deteriorate from normal walking to quadriplegia within a week to 10 days. Specific triggering viral infections can be influenza, CMV (cytomegalovirus), Epstein Barr virus (infectious mononucleosis), Zika virus, HIV and others. A recent worldwide survey[69] called IGOS (International GBS outcome study) spearheaded by neurologists in Rotterdam found interesting variations in the exact trigger. In Bangladesh, for example a common trigger was infectious diarrhea from a bacteria called Campylobacter jejuni, a common cause of food poisoning. While this infection also precedes GBS in North America and Europe it is less common than viral infections. GBS is described as "ascending" because longer nerves are damaged more easily. A person with GBS may experience tingling then numbness, sometimes painful, in the feet, then the legs, then the fingers, arms and trunk. If it progresses, all portions of the peripheral nervous system may be damaged. Because GBS often involves both sensory and motor nerves, weakness may start in the ankles, then at the knees, hips, hands and shoulders. In severe GBS, there may be complete paralysis including muscles of the face and eyes. Literally no muscle activity is present, creating a "locked in syndrome," a dramatic condition we will discuss below. Axons of nerves that supply respiratory muscles and breathing can also be involved. In the past this meant death but now it means admission to the ICU, intubation and ventilator support until you can breathe on your own again. GBS is characterized by complete loss of the deep tendon reflexes that in turn reappear during recovery. Improvement occurs gradually over days to weeks to months but in severe GBS permanent disability can result. More common are milder forms that nonetheless impair walking and work.

Why are all types of GBS not the same? This may depend how intense the autoimmune reaction is and its exact target. Fortunately most people do not develop GBS after food poisoning from infections such as Campylobacter jejuni, perhaps because subtle differences in the bacteria and self are enough for the immune system to tell the difference. The immune system may in effect be saying, "Do not go after these nerves because they belong here"! In most types of GBS, myelin bears the brunt of the immune attack, a demyelinating polyneuropathy. This is not to be confused with "demyelinating" problems in the CNS such as multiple sclerosis (MS). Major differences in the protein structure of PNS compared to CNS myelin trigger different kinds of autoimmune disorders. Other, somewhat less common types of GBS leave the myelin alone but damage the axons directly. It seems that Campylobacter infections associated with gastroenteritis are more likely to trigger an "axonal" rather than demyelinating type of GBS. Drs. Tom Feasby and Angelika Hahn, neurological mentors from the University of Western Ontario, first pointed out this distinction and I was privileged to be part of the seminal work.[70] Shortly thereafter, other cases of "axonal" GBS were recognized and a new name was applied: AMAN or AMSAN, acute motor (and sensory) axonal neuropathy. Drs. Guy McKann Jack Griffin, David Cornblath, Tony Ho and others from Johns Hopkins and China discovered an "epidemic" of axonal GBS in children from rural China.[71] Many cases were fatal. Preceding infections with Campylobacter jejuni may have triggered many of the China AMAN cases. Some GBS types are remarkably selective. To make the matter yet more complex, we also think that some severe instances of GBS damage both myelin and axons. The GBS may start out as "demyelinating" but involve such severe inflammation that nearby axons also get damaged, the "bystander" effect. It may explain why apparent demyelinating GBS, albeit severe, can be so slow to recover and leave permanent weakness. Think of a serious road accident where the people on the sidewalk are injured and you will understand the bystander effect.

The "Miller-Fisher Syndrome" (MFS) mentioned earlier, is an example of a GBS subtype and its eponym has persisted, with good justification! MFS is named after a famous Canadian neurologist Miller Fisher who happened to work at Harvard. Among his other contributions he described lacunar infarcts, a type of stroke, and a series of other strange neurological pain syndromes. The MFS involves only the eye movements, balance (ataxia) and deep tendon reflexes. Unlike most other forms of GBS, the exact molecule on the nerve that is targeted is known, called GQ1b ganglioside; elevated levels of antibodies directed toward it help with the diagnosis. Many Canadians have made remarkable contributions but have done so in other countries. This does remind us that the Canadian talent pool is deep and that further efforts and research support to foster success at home is sorely needed.

A careful neurological examination will usually identify the correct diagnosis when a patient presents to a health care provider with neuropathy symptoms. Electrophysiological tests of nerves that we have described earlier are critical in identifying causes of sensory loss or paralysis. These tests are, of course, normal in MS, whereas in GBS they not only identify the diagnosis but help classify the subtype and prognosis. It is quite important to know what kind of GBS has developed in a given person. Demyelination can be repaired by SCs within weeks, restoring conduction and reversing conduction block. Axon damage is difficult to repair, requiring the nerve to completely regrow from the site of the damage, akin to what we have described by axonotmesis. Given these stark difference, persons with "axonal" GBS recover much more slowly than those with the more common form of demyelinating GBS. If you have had axonal GBS you might be left with permanent wasting of the muscles supplied by the damaged nerves in your hands, arms and legs below the knees. The late Brian Langton, a Calgary friend and GBS survivor wrote about both his acute GBS and his recovery in two books: "A First Step. Understanding Guillain-Barré Syndrome" and "Guillain-Barré Syndrome, 5 years later" (written with other GBS survivors, including Patrick Hill and veterinarian Dr. Sarah Ondrich).[72,73] We lost Brian in 2013 from an unrelated illness but his account of GBS is part of his legacy. Brian's books well describe the course and slow recovery from severe GBS.

Only two established treatments for GBS are available at this time and both shorten the recovery time from GBS. One is plasma exchange (PLEX), a procedure that resembles hemodialysis, and that is thought to remove antibodies and other molecules that inflame the nerves. The other is intravenous gamma globulin (IVIG), a blood donor product. IVIG is thought to block harmful inflammatory molecules because it includes antibodies and other constituents from the donor pool that interrupt inflammation. Neither PLEX nor IVIG are "magic bullets" and neither result in a "Lazarus" effect in severe GBS patients. A "Lazarus" effect occurs when someone leaps from bed virtually cured after treatment, like the Biblical Lazarus who arose from the dead. We can only hope someday for these kinds of treatments, but applied of course while the victim is still alive! Despite the knowledge that GBS is an inflammatory disorder, prednisone, a well known anti-inflammatory drug used for the disorder, worsened the disorder when tested by clinical trial. Its use is now no longer recommended. This was a further lesson that careful clinical trials are essential in distinguishing between what we expect will help to what truly makes a difference during careful testing.

## CIDP

"Chronic inflammatory demyelinating polyneuropathy" (CIDP) is a complex name for an uncommon condition that resembles GBS but it is slow in onset or chronically waxes and wanes.[74] It is a demyelinating and autoimmune polyneuropathy without a preceding infection or known trigger. In most patients, its cause is unknown. Recently however, Luis Querol from Barcelona and his colleagues have discovered that some forms of CIDP have abnormal antibodies in the blood that attack proteins in the Node of Ranvier.[75–77] These antibodies are called anti-contactin, anti-neurofascin 155 and others. Recall how anchoring proteins and others at Nodes of Ranvier are essential to maintain intact myelin-axon connections. Unlike typical CIDP, these conditions do not respond to IVIG but fortunately do respond to a monoclonal antibody treatment, rituximab, that attacks antibody producing cells. We have a great deal to learn about these "nodal" polyneuropathies. Another cousin of CIDP has autoantibodies to one of the key proteins embedded within myelin called anti-myelin associated glycoprotein (anti-MAG). This disorder has very prominent demyelination that targets myelin segments further out along the nerve, called by some "distal acquired demyelinating symmetric polyneuropathy" or "DADS"! Mutifocal motor neuropathy (MFMN) is a condition in which several demyelinating autoimmune mononeuropathies target motor myelinated axons only. Some of these patients have autoantibodies to a ganglioside called anti-GM1. Finally Richard Lewis and Austin Sumner described multifocal acquired demyelinating sensory and motor neuropathy, similar to MFMN but also involving sensory axons and more easily remembered as the Lewis-Sumner syndrome. Its autoantibody is yet to be discovered.

Although at this time most CIDP patients do not have known autoantibodies, we believe that additional versions are likely. Typical CIDP remains a mystery. It causes weakness, numbness and inability to walk. Fortunately and unlike GBS, CIDP almost never requires an ICU or ventilator. Nonetheless, CIDP can be quite disabling and painful. Nerve conduction tests show demyelination, not unlike that identified in GBS. Over time, nerves can enlarge and thicken because their SCs begin to multiply, or proliferate in the nerve, in a misguided attempt to restore myelin. The endoneurium develops whorls of Schwann and other cells surrounding axons, forming interesting structures under the microscope called "onion bulbs," resembling layers of an onion. Like leprosy, the nerves in CIDP can become so enlarged that they are visible under the skin or on MRI studies. Also unlike GBS, CIDP can be treated with prednisone that dampens nerve inflammation. However prednisone has many longer term side effects that include weight gain, osteoporosis, diabetes mellitus, hypertension, cataracts and others! While prednisone has not been associated with a "Lazarus" awakening, it can have remarkable and major benefits in CIDP; sometimes life long treatment may be required. Over the past 10 years, IVIG, while much more expensive than prednisone, has been recognized to be better tolerated and can be infused monthly to maintain function in persons with CIDP. With ongoing treatment started early, most people with CIDP can function well.

What we have learned from CIDP is that a polyneuropathy thought "simply" to attack myelin often does much more. Bystander axons can also become damaged during acute inflammation (GBS) or chronic lower grade inflammation (CIDP) complicating recovery. The intimate relationship between axons, SCs and myelin means that damaging one very often leads to damage of the other. Currently our treatments are limited to arresting the inflammation in CIDP or GBS to allow the nerves to repair or regenerate themselves. To recover, axons are required to reconstitute their myelin, erect normal nodal architecture and restore normal conduction. More targeted treatments to prevent axon damage or encourage more vigorous regeneration is not yet available in the clinic or hospital. More research is needed and more support for it!

## Bolton's neuropathy (critical illness polyneuropathy)

In the early 1980s, Dr. Charles Bolton, a colleague and mentor, described a condition never before documented. As a neurologist at Victoria Hospital in London, Canada he was repeatedly asked to evaluate patients in the ICU, not a common request. The ICU or CCTU (Critical Care Trauma Unit) at this hospital had an aggressive support program for its high patient volumes and was overseen by a highly regarded intensive care specialist, William Sibbald (1946–2006). Back in the 1970s and 1980s, Neurology was mainly a diagnostic specialty and its visits to ICUs were largely confined to catastrophic brain injury from cardiac arrest, trauma or infection. Neurologists have changed in the last three decades, are of diverse backgrounds and gender, and are intensive and empathetic in their involvement and treatment. The advent of brain and spinal cord imaging in the 1980s also changed the role of Neurology to focus on treatment, not strictly localization. Despite this "sea change," fine and complex peripheral nerves are not well seen and studied by current imaging techniques and the older skills of localizing remain very important.

Charles Bolton was not an ICU-reluctant neurologist. He was open minded and curious while rigorous in his approach. William Sibbald asked him to sort out why so many CCTU (ICU) patients developed generalized muscle weakness, some so severe that they could not walk or breathe and were dependent on a ventilator. In the ICU, as a person recovers from a devastating illness requiring respiratory support from a ventilator, the usual approach is to slowly "wean" them from using it. At the outset of their illness, a ventilator may be lifesaving if there is serious disease of the chest or lungs but as these are treated and recover, further ventilation should not be required. Most patients are glad to escape to a regular hospital ward when successfully weaned, then home. William Sibbald's patients seemed very challenging to wean despite improvement in their condition and resolution of multiple organ failure. Clinicians had seen something like this before, but came up with ideas such as "muscle fatigue" to explain an apparent inability of respiratory muscles and the diaphragm to resume normal and independent breathing. Charles Bolton discovered that these patients often had weakness in their limbs, sometimes complete paralysis and loss of their deep tendon reflexes. By using electrophysiological testing, it became evident that they had an unexplained axonal polyneuropathy. In patients with severe involvement, the neuropathy involved nerves supplying intercostal (rib cage) muscles, the diaphragm and limb muscles but curiously sparing face and eye muscles. Many did not survive the CCTU and several of these patients were studied by autopsy.

Bolton described his first four cases of this condition at the American Academy of Neurology meeting and in a paper entitled "Critically ill polyneuropathy" in 1984.[78] Many neurologists were skeptical although their direct experience venturing into ICUs and carrying out electrophysiological studies there was limited. Those offering resistance to the concept of a newly identified polyneuropathy suggested that Bolton had missed the diagnosis, misclassifying GBS for example, or another rare toxic form of nerve damage. Steadfast belief and Bolton's careful observations eventually won the day and ICUs around the world recognized that they had been witnessing identical problems with their patients. Today "critical illness polyneuropathy" (CIP) is an established diagnosis. In recognition of the work of a rigorous and singular physician investigator, CIP is also referred to as "Bolton's neuropathy." As a resident in the mid 1980s I had the privilege of working with Charles Bolton and characterizing detailed autopsy findings on 9 of the first 17 of these patients.[79] Despite the time that has gone by and the acceptance of it as a common ICU problem, the precipitating cause for Bolton's neuropathy remains unclear. It emerges in persons originally admitted to ICUs for non-neurological problems such as trauma or cardiac disease. While Bolton's neuropathy does prolong the stay of persons admitted to ICUs and prolongs their recovery, the condition resolves itself in the manner and time-frame of widespread axonal damage. Since the polyneuropathy requires axons to regrow and reconnect, recovery is slow, sometimes incomplete. Many patients do not survive their underlying illness and only 60% or fewer are successfully discharged from the ICU. Since Bolton's neuropathy was initially described we have recognized that ICU patients can also develop direct damage to their muscles, termed "myopathy" (myo = muscle). Given these additional findings, the whole concept of patient weakness in the ICU setting (ICUAW-ICU acquired weakness) is more fully understood and treated. We have knowledge of how both CIP and myopathy can develop and even coexist. In the context of widespread multiple organ failure and severe infection known as sepsis, it is no surprise that both muscles and nerves are targeted. Moreover, the brain can also be targeted in ICU patients, causing confusion and delerium termed "septic encephalopathy." Finally, weakness and damaged muscles have also be been linked to specific muscle relaxing medications administered to ICU patients. These agents were designed to paralyze patients in order to facilitate their support on ventilation. However with prolonged use of nondepolarizing muscle blocking agents such as pancuronium and vecuronium, muscle damage and blockade of neuromuscular junctions may develop. Overall muscle damage in the ICU is now called critical illness myopathy, although other entertaining names were originally used such as "floppy person syndrome"! ICUs are expensive and challenging places to stay, given the presence of tubes and IVs and difficult infections. Since the 1980s, intensive care specialists and their colleagues have also learned more about how long to treat people with overwhelming disorders or when ongoing treatment attempts might also be futile without hope for recovery. Aggressive "no holds barred" treatment is appropriate for persons with reversible disease who consent to this kind of care. For others, with serious and multiple complicating conditions, there can come a time when more humane but less aggressive treatment is requested. We do recognize however that the neuromuscular complications of an ICU stay such as Bolton's neuropathy and critical illness myopathy can fully recover and that patients may need additional time to do so.

## Other "acquired" polyneuropathies

Acquired polyneuropathies, some of which are addressed here, are distinct from the inherited polyneuropathies. Fifty percent or more of chronic polyneuropathies are "idiopathic" which simply means that their cause is unknown. The term "idiopathic" however can be disturbing to persons suffering from it. Many patients with idiopathic polyneuropathies become despondent because of the lack of understanding or treatment for their neuropathies. Eventually we will discover that some of these are inherited, perhaps recessively, meaning that the DNA abnormality arises from a recessive gene without clinical parental involvement. Some idiopathic polyneuropathies may develop from low grade autoimmune disease and others from undiagnosed underlying systemic disorders that can involve the nerves among other organs. A small percentage of "idiopathic" polyneuropathies go on to have CIDP, vasculitis or rarely an unrecognized cancer. Polyneuropathies related to cancer are called "paraneoplastic." "Neoplastic" refers to a malignant process. "Paraneoplastic" neuropathies do not arise from direct tumor cell nerve invasion but rather as an autoimmune reaction. This might be similar to "molecular mimicry" we discussed under GBS, when the body mistakes a nerve protein for something foreign to attack. In this case, it may be that the cancer proteins, newly released into the bloodstream by diseased cells, cause a misplaced immune reaction. A new sensory polyneuropathy in an apparent healthy middle aged person, for example, that appears and progresses over a few weeks to months raises suspicion of a paraneoplastic cause. Such persons warrant having a thorough check over for breast, colon and lung cancer.

Treatments for cancer are also a common cause of polyneuropathy, most often axonal in type and particularly targeting sensory neurons. Some of the forms of chemotherapy causing neuropathy include vincristine and other vinca alkaloids, cisplatin and other platinum therapies, taxanes such as paclitaxel and docetaxel, bortezomib, thalidomide, suramin and others but the list here is incomplete (see Refs. 80,81). These agents damage nuclear and mitochondrial DNA of neurons (platins), microtubules that are required for axoplasmic transport (taxanes and vincas), or alter how nerve proteins are handled by cellular structures called the proteasome (bortezomib). Rarely some of the new biological treatments such as infliximab or immune checkpoint inhibitor agents for cancer have been linked to demyelinating polyneuropathy.[82,83]

Low thyroid function, kidney failure and other rare metabolic disorders are also causes of acquired polyneuropathy. HIV infection preferentially damages small unmyelinated fibers. Patients have prominent pain with loss of sensation to pain and temperature. Interestingly some of the treatments for HIV also can cause polyneuropathy, sometimes aggravating the existing problems of patients.[84]

## Polyneuropathies you inherit

Beyond diabetes mellitus, the most common polyneuropathies are inherited. Among the more serious and devastating types are those associated with "amyloidosis." Amyloidosis was first described by pathologists as a collection or "blob" of amorphous protein deposited in tissues such as the skin, heart or nerve. It had not been clear if amyloid caused damage to the tissue it was identified in or instead was a byproduct. Amyloid possesses a very distinctive apple green color staining when highlighted by a polarizing filter microscope and a "Congo Red" stain. We now know that amyloid deposits are caused by the aggregation of sheets and chains of amassed protein collections. A variety of abnormal proteins can cause an amyloid deposit. Amyloid polyneuropathy develops from amyloid staining material within nerves from a number of distinct, sometimes mutated proteins. Yet another type of amyloidosis with a differing underlying protein accumulates in the brain (only) of persons with Alzheimer's disease. Amyloid in nerve can arise from deposition of immune proteins complicating a bone marrow disorder as lymphoma or multiple myeloma. If the tumor is cured, the amyloid can eventually dissipate and the neuropathy improve. In practice however, these patients are often very ill and many do not recover. Unfortunately amyloid can also target the heart, damaging its muscle and destroying the conduction system it uses to sustain its rhythms. Cardiac autonomic axons are also damaged. Amyloid polyneuropathy preferentially damages small sensory and autonomic axons first and persons may injure their digits by burns or other injuries without realizing it. These injuries are often complicated by infections, and nonhealing ulcers that eventually require amputations. Pain is often severe. Since autonomic fibers are damaged, there may be intestinal symptoms such as diarrhea and constipation. As the disorder progresses, motor axons and larger sensory axons are targeted.

Inherited amyloid polyneuropathy occurs most often from a mutation in a protein called transthyretin. Patients with these mutations were originally thought to be of Portuguese ethnicity but the condition is now recognized worldwide.[85,86] The important news here is the emergence of new treatment options in the past five years. These have included drugs that prevent the proteins from folding abnormally into amyloid, allowing the body to clear existing deposits.[87] Other treatments inhibit the mRNA from forming proteins that could aggregate into amyloid.[88,89] An older treatment has been liver transplantation, since the amyloid forming proteins are made in the liver. Transplantation is obviously a large undertaking and must be carried out before the polyneuropathy is severe and irreversible. Proceeding with this approach is obviously not an easy decision but may slowly be supplanted by the new targeted approaches discussed above. We do not have a full picture of how well the new treatments will work, but it is an exciting positive change. It should now be clear that neurologists no longer only diagnose!

The most common inherited polyneuropathies are called CMTs for short.[90,91] CMTs are actually the most common genetic abnormality in humans! CMT stands for Charcot-Marie-Tooth disease, named after some very ancient neurologists, Jean Charcot (1825–1893; the most famous), Pierre Marie (1853–1940; his student) and Howard Henry Tooth (1856–1925; a little more obscure). Charcot and Marie were famous French neurologists at the iconic Salpêtrière Hospital in Paris whereas Tooth was British, appointed to hospitals in London. This British-French dynamic may have complicated the "naming" in those days, but we are left with it. Incidentally Sigmund Freud was also a neurologist and a student of Charcot but regrettably abandoned neurology for something else. The rest is history. Jules Dejerine (1849–1917), also a French neurologist who practiced a little later at the Salpêtrière, described a variant CMT in younger onset patients called Dejerine-Sottas disease with arguably the most severe version of inherited demyelination known.[92]

A few decades ago Dr. Peter Dyck from the Mayo Clinic, Rochester, Minnesota, tried to eliminate these CMT eponyms. He termed them "Hereditary Motor and Sensory Neuropathies" or HMSNs of various numbered types. His detailed work and characterizations were a major leap forward in the field.[93] HMSN was never a household word and was cumbersome even for neurologists. Molecular biologists and geneticists thereafter identified the abnormal DNA of genes that cause these conditions and reverted to the shorter, functional "CMT." This has stuck but some of the numbering is still applied. For example, CMT type 1a is the most common type. It is caused by a mutation of the protein that resides in myelin called PMP22. As you might predict, its mutations (actually a duplication of the gene) disrupt the normal structure of myelin. Macrophages detect the abnormality and start to remove myelin. A demyelinating polyneuropathy is the consequence. Like CIDP, SCs continually try to replace their damaged myelin and form "onion bulbs" we described earlier: whorls of SCs and other cells that surround a demyelinated axon. Nerves containing onion bulbs are enlarged and form thick ropy structures palpable under the skin as described with CIDP or leprosy. Electrophysiological studies are highly abnormal. Recall that normal conduction velocity in adult human nerves ranges from 40 m/s in the lower limbs to 60 m/s in the uppers. Persons with a CMT1a mutation may have conduction in the range of 10–35 m/s, a dramatic difference. However severe slowing of conduction itself does not cause disability. Conduction block, linked to weakness in GBS or CIDP, is not a feature of CMTs. In addition to demyelination however, CMT patients sustain severe loss of axons especially in tibial and peroneal nerves that plantarflex and dorsiflex the feet respectively. Over time, the axons also disappear in other territories including those supplying hand muscles. The denervated muscles develop weakness and wasting, a consequence of lost motor axons. Persons with CMT also have high arches in their feet, an unexplained feature that may develop in utero from abnormal muscle innervation.

At this stage the reader might stop because all of this sounds contradictory. For example CMT1a is called a demyelinating polyneuropathy. However axons also disappear and contribute to disability and foot deformities, the more serious consequences of CMT. The answer is that nerves with chronic abnormalities of SCs and myelin eventually also lose their axons. If SCs are programmed to have an abnormal myelin gene, the axon is aware of the difficulty and does not fare as well. SC axon interactions are complex and essential as you have heard several times earlier in this book.

CMTs are numbered in broad categories then divided into whether they exhibit prominent demyelination or exclusive loss of axons without demyelination. While demyelinating CMT1s have later axon loss, CMT2s have axon loss first and foremost. The nerves are not enlarged and ropy and nerve conduction velocity is not severely reduced. However, in CMT2 there is foot drop, wasting of the hands and legs and high arches! All CMTs seem to have these features in varying degrees. Some start earlier in life, others later. A quick note about wasted legs. This is a typical feature of CMT called "inverted Champagne bottle legs." The tops are wider, the lower legs narrower as a result of muscle wasting. The description does sound Parisienne and may have originated from Charcot or Marie.

Despite the fact that CMT1a is the most common type of CMT, many others exist depending on the exact genetic mutation. Genetic studies here have yielded new insights into nerve biology. For example, some types of CMT2 arise because of a mutation in a protein called "mitofusin." Mitofusin is important for how nerve mitochondria connect. Mutated mitofusin might cause polyneuropathy because of defective mitochondria and energy failure in the axon.

Some CMTs are inherited as an autosomal dominant trait. This means that a son or daughter has a 50% chance of inheriting the abnormal gene if their parent has CMT. Others are recessive (e.g., CMT4s) so that neither parent has obvious CMT but their children can develop it. Still others are X linked so that only boys develop it. Recall that since females are XX, only one of their X chromosomes can transmit CMT while the other compensates. In males, there is only one X (XY) so that if they inherit their X from their mother with a mutation in that chromosome, they will develop CMTX. Discoveries over the biology of CMTs are ongoing. Many are presented at annual Peripheral Nerve Society meetings that bring together key scientists working on them and neurologists who see them as patients. The number of independent mutations causing a CMT syndrome now exceeds over 90! However routine testing for only a few have been available, pending wider use of whole exome or whole genome DNA sequencing.[90] Many forms of CMT are discovered in childhood but some do not emerge until adulthood as an unexplained "idiopathic" polyneuropathy. Wider availability of genetic testing for all described CMT mutations will eventually allow routine testing of older persons with unexplained polyeuropathies.

We do not have therapy available that will reverse CMT. Fortunately, it is not a life threatening disorder excepting one rare form with vocal cord paralysis. An old clinical rule about CMTs provided by neurologists was that it does not shorten life span and rarely leads to the need for a wheelchair. The first is generally true but some severe types might require a wheelchair in later life. CMTs are slowly progressive over time. Over decades CMT persons can have gradually more weakness in the hands and feet. Some find help from using what is called an ankle foot orthosis, simply a brace that prevents footdrop. Occasional CMT patients have pain and there are non-narcotic pain treatments that can help. Few clinical trials have been conducted. For example, while Vitamin C helps SCs form new myelin a trial of it in CMT patients offered little benefit. We require more "guided missile" treatments such as those being developed for amyloid polyneuropathy. This might include treatments to replace mutated proteins with normal ones. I only speak for myself, but at the time of writing this (2019) I continue to see CMT patients with advanced neuropathy, or even early neuropathy to whom we have very little to offer.

CHAPTER 6

# Locked in

What would it be like to lose all of your nerves-motor, sensory and autonomic? The more important question is whether you might survive this. Even in an ICU setting with full ventilator support for your lungs, the likelihood of survival in this scenario would be reduced. Critical body functions become erratic in the absence of autonomic nerve function. For example there may be sudden changes in cardiac rhythm, heart rate or blood pressure that are life threatening. Fortunately most neuropathies, even if very severe do not result in this kind of catastrophe. Polyneuropathies only rarely progress to complete paralysis. For example, a patient with very poorly controlled diabetes may develop loss of sensation as far up as to the shoulders and thighs, but they continue to have muscle power.

The "locked in" syndrome is a rare neurological event. It refers to an inability to communicate with the outside world by speech, movement or other means. The scenario is terrifying to think about but some have survived to tell their story. When a person is comatose, they are unarousable and unresponsive. This means that they cannot be "awoken" and do not give any hint that they can respond. However, how does someone respond or appear to rouse if they are experiencing complete paralysis? Neurologists are trained to sort this out but it takes careful assessment. For example, many neurological disorders from meningitis to bilateral strokes to seizures cause loss of consciousness. However very few cause the locked in syndrome, a situation in which a person is incapable of any movement while being awake and aware.

There are several causes of this syndrome. One is a very specific stroke. Recall that strokes arise from death or "ischemia" of a part of the brain from blockage of its blood supply. Rare strokes that involve the brainstem, from damage of the basilar artery that supplies this part of the brain can cause a locked in syndrome. The brainstem is a critical structure because it connects the cerebral hemispheres to the spinal cord and thus it channels all of our movement and sensation. It consists of an upper portion called the midbrain, a middle portion called the pons and a lower portion called

*Our Wired Nerves*
https://doi.org/10.1016/B978-0-12-821487-9.00006-4

the medulla. A stroke that damages the middle of the pons interrupts the communication between the cortex of the brain and spinal cord. A locked in syndrome results. A pontine stroke interrupts both movement and sensation and if the damage also involves the posterior portion, some eye movements are also impaired. In this difficult scenario, reaction of the pupils is retained, since light pupillary reflexes are activated by a different pathway in the spared upper brainstem. Given this, an important clue that someone might be locked in is the presence of reactive pupils to light, a preserved pupillary reflex or some residual eye movements. Thus, while the person is unable to speak, they might, for example, be able to respond yes or know by blinking or rolling their eyes to a limited extent. A recent book and film titled the "The Diving Bell and the Butterfly" by JD Bauby, a French journalist, describes his experiences as a patient locked in from a stroke but who managed some recovery. It is a very compelling story to read or see on film.

There are other causes of "locked in" syndrome. Like those caused by brainstem strokes, they may not be reversible. A rare condition called central pontine myelinolysis, referring to myelin damage in the middle of the pons, develops because of rapid changes in blood sodium. These changes may develop because of a metabolic disorder or may be related to in hospital treatment of low, sometimes life threatening levels of blood sodium. Central pontine myelinolysis can recover, but slowly. Another cause may be secondary to a large zone of inflammation from multiple sclerosis known as a plaque. Rare brain tumors, including a devastating version known as brainstem glioma, will cause locked in syndrome, sometimes in children. The prognosis for this condition is very poor at this time. While many tumors can be fully or partially removed by neurosurgeons, it is impossible to resect a brainstem glioma without causing further damage to the very sensitive structures of the brainstem. Severe endstage ALS, or motor neuron disease can also be associated with a locked in syndrome. This is a result of loss of most of the motor neurons that control movement although sensation is preserved. The physicist Stephen Hawking suffered from this condition, communicating with minimal cheek muscle movements to a computer. While uncommon now, victims of poliomyelitis suffered from such severe paralysis that they required ventilators. These were called "iron lungs" in the early part of the 20th century. While these patients retained sensation, motor weakness was severe and widespread. Most, but not all recovered.

The "locked in" state can also occur because of severe Guillain-Barré syndrome (GBS). While full "locked in" is uncommon in GBS, lesser variants of it are quite common. Severe demyelination of nerves, with or without axonal damage can interrupt the function of all motor nerves. Limb nerves, facial nerves and nerves that control eye movements develop conduction block or in some instances, axonal degeneration. Rarely, even the pupils may fail to react because of damage to autonomic nerves. It might be very challenging in establishing if a GBS patient is "locked in" or comatose. Knowing how the problem evolved and establishing the diagnosis is crucial. Since most instances of GBS are associated with demyelination, they are most often reversible; this is very important news for the sufferer and their family. Caregivers require reminding that what they say around the patient may be heard and interpreted by locked in persons. Many find the ordeal harrowing but like other locked in patients they may have interesting dreams, nightmares, hallucinations and fantasies during their paralysis. One of the most interesting examples of this is a book entitled "A First Step. Understanding Guillain-Barré Syndrome" published in 2002 by Brian Langton, a friend described earlier.[73] Brian had a severe variant of GBS that rendered him ICU bound for nearly 7 months. He was paralyzed and respirator dependent. Recovery eventually developed as he put his heart and soul into working hard with therapists to recover lost function. Four years later he was independent in activities of daily living but had severe weakness and wasting of his hands and feet. In his book Brian describes over twenty short anectdotes about his thinking and dreams while paralyzed. Some are quite amusing. One example is his attempting a long drive while navigating packs of aggressive wild dogs along the highway that are attacking him and other travelers. During the dream events he has pain in his limbs that he attributes to the bites from the attack dogs. These eventually awaken him back in the ICU. The pain was likely quite real, a product of his damaged nerves, but built into another story by his partly disconnected brain. Brian may have been interpreting neuropathic pain from GBS as bites of wild dogs. In another dream he rescues a train from a falling bomb. The story ends by someone congratulating Brian with the very appropriate words, "You have guts!" This was a very accurate appraisal. He survived his ordeal, wrote two books about it and retained an extraordinary degree of optimism about his experience. I had the fortune to meet Brian several years after his illness when he was more or less independent but also relentlessly cheerful. The interesting part of these stories is how they relate to his experiences and his overall life story. It seems that the loss of nerve input and output, even for a temporary period during life profoundly alters thinking and introspection. I can think of no better example of how our nerves and thinking are part of the same mix, carefully interwoven.

Even less common than a "locked in" paralytic syndrome is what is called complete "deafferentation." This is simply a term that means complete loss of sensation without motor damage. Oliver Sacks in his book "The Man who mistook his wife for a hat" highlights this rare problem.[94] In his chapter "The Disembodied Lady" he describes a person entirely impaired because of the lack of sensation, despite her ability to send messages to her muscles through motor nerves. Complete sensory loss from peripheral neuropathy most often occurs because of an autoimmune disorder, but is almost never associated with GBS. For unclear reasons, GBS appears to target motor axons more frequently and with greater severity than sensory axons. Thus deafferentation syndromes are different kinds of polyneuropathies than GBS. While Sacks explores what it is like to be deafferented, he does not describe how it might come about. In one patient seen by the author with this syndrome, the patient developed a very unusual autoimmune condition with inappropriate antibodies against a key protein in her sensory nerves known as GW182. GW bodies are recently discovered proteins, by a colleague in Calgary, Dr. Marvin Fritzler, that control how RNAs are degraded.[95,96] You will recall that RNAs serve as messengers or templates from DNA to protein making machinery. We are learning a great deal about RNA function and nerves but the story is incomplete. In a series of newly understood inherited neuropathies, defects of RNA function from one or more mutations can cause dramatic changes to nerves, sometimes damaging. Why an antibody to GW182, however, would render such a severe condition as deafferentation is not clear at this time. An important aspect of deafferentation is that it dramatically impairs walking. At first glance this does not make sense, since walking would seem to depend only on motor nerves. These are the nerves that signal muscles to contract and move our limbs. Missing, however, is the incoming data from muscle spindles and tendon sensory afferents as to where those limbs end up. Did the limb travel far enough, touch ground or avoid an obstacle? My deafferented patient with strange anti-GW antibodies was completely unable to either stand or walk.

# CHAPTER 7

# The disrupted connectome and pain

There is a kind of perverse justice when nerves are damaged. One might expect loss of sensation or movement. However, nerves express their unhappiness over damage in a much more problematic way. Injured nerves generate pain. The pain is not only common after nerve damage, it can be very intense and unremitting. In fact, the pain problem is so serious and common when our connectome of nerves is disrupted that it has a category of its own, "neuropathic pain."

Neuropathic pain has distinctive qualities. In fact, these qualities can help tell if the pain is from a damaged nerve versus some form of joint or tendon injury. They are called positive symptoms but not because they are positive for people! This simply means that instead of the nerves acting as passive transmitters of sensation, they begin to act up on their own. They are hyperexcitable and they generate impulses when they should not. We discussed some aspects of this earlier in this book. One of the most common sensory symptoms is tingling or prickling or a "gone asleep" sensation. These sensations are not the result of normal signals arising from the body that need to be transmitted. They are aberrant, or inappropriate "positive" sensory symptoms that may indicate damage to a nerve. It is interesting that the tingling may often occur in the exact distribution as the normal sensation supplied by a given nerve. For example, the ulnar nerve supplies sensation to the inner half of the hand, the small finger and the inside half of the ring finger. When you bang or bruise your ulnar nerve around the elbow, tingling can occur. In common terms, people report that they hit their "funnybone"! The tingling may or may not be painful but can be felt in the identical distribution-inside of the hand, small finger and part of the ring finger. Sometimes the tingling may be hard to localize to one area. It may feel as though the whole hand or leg is tingling when only one nerve is responsible. This is an interesting property of our nervous system and you can blame the brain and CNS for it! Telling our conscious brain where an injured nerve is firing is filtered information. If information like this arrives

*Our Wired Nerves*
https://doi.org/10.1016/B978-0-12-821487-9.00007-6

as a barrage of discharges from an injury, the brain might not interpret it as arising from a single nerve territory. It may interpret the sensation as arising from a limb, of unknown significance! Thus one might feel as if the entire arm, hand or leg has "gone asleep." This is very common in carpal tunnel syndrome as we discussed above.

Sometimes tingling does not represent a nerve injury at all. For example we have discussed persons with very low calcium levels in their blood that may experience the same sensation. Many persons also mistakenly worry that tingling means "poor circulation." In people who indeed have "bad circulation" such as narrowed arteries, tingling arises from a compromised blood supply and oxygenation to nerves, known as ischemia. Nerves with ischemia generate abnormal APs with positive symptoms like tingling. If ischemia is prolonged and severe enough, later death of the axon ensues. Nerves are actually more difficult to make ischemic than the brain, but that is another story. The question is why a borderline ischemic nerve generates positive signals! Energy is required to maintain the delicate electrical balance of the nerve, pumping some charged ions like sodium out of the nerves to prevent their accumulation. When the energy supply is compromised by ischemia, axons lose this critical electrical balance and are depolarized, a property we have discussed earlier. Inappropriately depolarized nerves, also called "ectopic" impulse generators, send off discharges that have no relationship to normal sensation and do not arise from normal sensory receptors.

Other painful "positive" sensory symptoms arise from damaged nerves (Fig. 32). Together they form a cauldron of "neuropathic pain" symptoms. Specific descriptions in English are: tingling, prickling, "pins and needles," burning, tightness or constriction, electrical sensations. Dr. Peter Dyck told me that Minnesota persons prefer to say "prickling" rather than tingling but I have not been able to verify this. In Alberta, both terms are common! Other kinds of descriptions of pain can accompany the above. Many also relay "deep aching" pain, a less specific and not well localized descriptor. Damage to muscles or joints, or "musculoskeletal" pain may be more likely to be deep and aching. This does not imply that nerves are not involved in sending this odd and unpleasant sensation to the brain! They are heavily involved because activated nerve endings in a damaged joint or muscle send the signal to our brain that there is a problem. Our nerves are sensors of what is transpiring in any given tissue in which they reside. This is an important distinction. In pain that is not neuropathic, such as musculoskeletal pain, axons are not damaged but are simply activated and doing what they

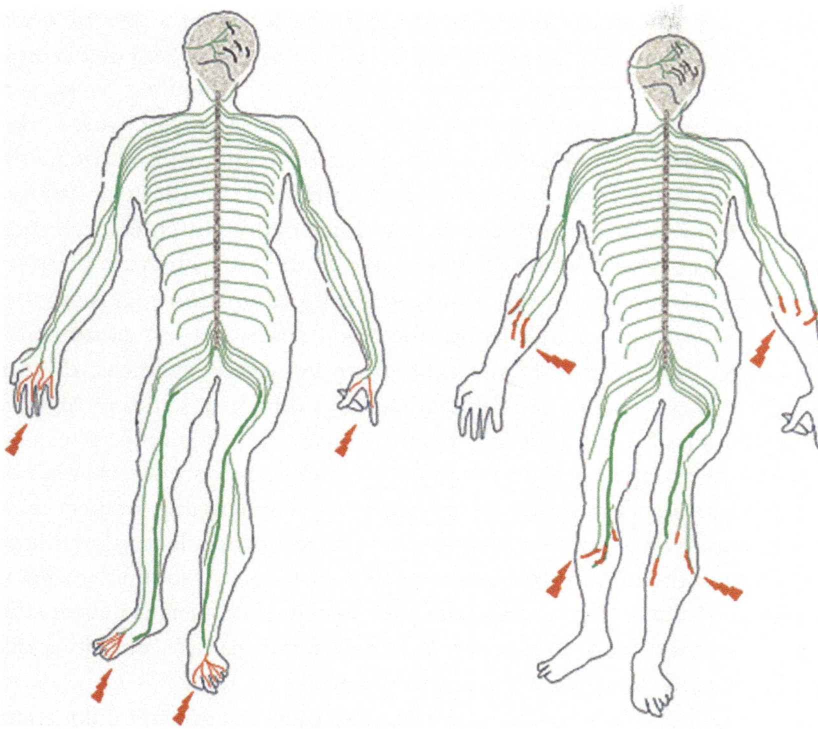

**Fig. 32 Diagrams illustrating sources of neuropathic pain in an early (left) and more advanced polyneuropathy. Pain also arises from sensitization of ganglia and the CNS (see text).**

are meant to do-signal pain. Neuropathic pain requires specific damage to the nerves.

Differentiating neuropathic and non-neuropathic pain can understandably be challenging. An official definition of neuropathic pain used by researchers studying pain or its treatment is[97]: "the presence of 'neuropathic' symptoms and evidence of damage to the somatosensory system as identified by a neurological examination or test." Here is an example of how the differentiation can be difficult. Low back pain is very common in adults and musculoskeletal problems are the main culprits. They include osteoarthritis in the back, bony spurs called osteophytes, muscle damage, or damage to facet joints in the back. The pain is typically deep, aching in the back or buttocks but it does not travel into the legs below the knees. It should not be associated with any tingling, numbness or other neuropathic symptoms. If there happen to be additional neuropathic features, this may change how we think about the cause of low back pain. If the pain is in a specific nerve territory, then it may be associated with nerve root damage. For example a disc protrusion in the back between the 5th Lumbar and 1st Sacral vertebrae may compress the first sacral (S1) nerve root. This can cause severe neuropathic pain that radiates (travels) down from the back, through the buttock and into the leg, and especially into the sole of the foot. Persons with an S1 nerve root compression describe tingling, tightness, electrical sensations and perhaps numbness in the sole of the foot. The pain territory fits into the S1 geography and a neurological examination may show evidence of S1 damage such as loss of the ankle reflex and loss of sensation, when tested, on the sole of the foot. In more severe instances, there is weakness and wasting of the calf muscles and difficulty standing on toes. One important caveat is that insufficient damage to nerve root axons may be associated with pain but without definite numbness or reflex changes on the neurological examination. In this instance, it is difficult to detect axon damage despite the suspicion that the pain is neuropathic. It may require an MRI to show the disc protrusion or an EMG to identify subtle damage. In our case of S1 nerve root damage, the pain would be defined as neuropathic if the symptoms fit into the correct nerve root territory and a corroborating MRI confirmed the cause if a disc was present.

More than one type of pain may coexist, such as both musculoskeletal and neuropathic pain. This is yet more challenging and unfortunately common. Back and other types of pain are so common in adults that the use of opioids for relief has fostered a crisis in North America. The sad part of this story is that while opioids can provide effective pain relief, their overwhelming risks of causing addiction have become a massive health care problem.

Alloynia is an important and interesting property of neuropathic pain. Most of us have experienced a heightened response to a light or innocuous touch that should not be painful but is unexpectedly uncomfortable. Allodynia is the perception that an otherwise innocuous sensory stimulus is painful. It is common to experience this nearby a healing wound where a touch stimulus can be exquisitely tender. For the same reason, persons with polyneuropathy may not be able to tolerate bedcovers touching their feet. Related to allodynia is another neuropathic sensation called "hyperalgesia," in which a very mildly painful sensation is dramatically ramped up in its intensity. Thus, a neurologist may be testing a person for their sensation with a dull safety pin tip. The patient reacts strongly, even inappropriately so, because their perception of the sensation is magnified. The reasons for these unpleasant symptoms are worth discussing.

Neuropathic pain is most often described when nerves are injured. An iconic neurologist during the time of the Civil War in the United States, Dr. Weir Mitchell (1829–1924), documented original descriptions of neuropathic pain due to nerve injury.[98] He described soldiers recovering from musket wounds and other injuries. Many had damage from a bullet partly or completely transecting the nerve. Bullets are associated with a "wave" of tissue damage around them that is much larger than their actual size. Thus a musket bullet might pass through the leg and miss the sciatic nerve by a centimeter, but its accompanying wave of damage is severe. A sciatic neuropathy results despite the fact that it was not in the direct pathway of the bullet, but instead a "bystander" injury. If the bullet path is further afield it may distort the tissues around the nerve and cause neurapraxia with local demyelination. These injuries also cause inflammation and swelling, injurious to the nerve, that arise later, after hours to days. Some may disrupt the arteries and veins that supply the nerves, causing ischemia. Weir Mitchell made interesting observations, such as the presence of allodynia, or how neuropathic pain might be relieved by soaking the limb in cold water! This is an instance of "Don't try this at home," especially if you are a diabetic with polyneuropathy. Soaking of limbs with nerve damage may indeed temporarily alleviate pain but the downside is that ongoing soaking damages the skin cells. Between skin cell damage and loss of sensation, chronic foot ulcers may be more likely to develop.

Thus, while we may regard nerve damage and neuropathic pain as separate entities from the CNS, keep in mind that the human nervous system is normally fully connected. This has implications for our perception of pain since the full neuraxis (nerves, spinal cord and brain) usually participates in chronic pain. In addition, CNS lesions themselves can cause troubling and problematic pain resembling that of nerve injury. The term "neuropathic pain" originated from peripheral nerve injury symptoms but is now widely adopted for central nervous system pain as well.

The spinal cord, especially its dorsal horn, can be responsible for CNS "neuropathic" pain. Recall that the dorsal horn at the back of the spinal cord gray matter is a relay center for sensory signals. When peripheral nerve fibers enter the spinal cord, they are no longer accompanied by SCs. CNS glia cells now support and provide myelin for the fibers. Nerve entry ports to the spinal cord travel through a door or portal called the "dorsal root entry zone (DREZ)." Why is this important? One of the most severe, although largely historical neurological conditions directly targets the DREZ. Before modern antibiotic treatments, persons with untreated syphilis developed involvement of the nervous system from invasion of the syphilis spirochaete, a development known as tertiary syphilis. One of the most characteristic syndromes of tertiary syphilis is a condition called tabes dorsalis, associated with severe neuropathic pain. Tabes dorsalis is a unique and highly localized inflammation of the DREZ. Patients with the condition experienced loss of sensation and characteristic pain syndromes; for example as if nerves had been "pulled out and plucked like a violin," "positive" sensory electrical like sensations. Fortunately tabes dorsalis is rare now given the availability of antimicrobial drugs. However, other disorders of the spinal dorsal cord such as MS or tumors can cause severe neuropathic pain.

The thalamus is a relay station in the brain that sends signals from the nerves and spinal cord to the cortex of the brain. A thalamic lesion caused by a stroke, tumor or MS can alter thalamic circuitry, generating inappropriate symptoms of pain felt on the opposite side of the body. Recall that pain pathways cross over the midline. Thalamic pain can be intense and severe with burning, tingling and allodynia involving the contralateral body, a hemicorpus (half of the body) pain syndrome. The pain geography cannot be ascribed to a single nerve or plexus.

Chronic neuropathic pain arising from a single damaged nerve recruits participation and remodeling of the entire pain sensing pathway over time. Rewiring may involve the cortex, the thalamus, the dorsal horn of the spinal cord, the dorsal root ganglia and the injured nerve itself. Damaged nerves send inappropriate discharges because their electrical properties are altered by the injury. These discharge sites have been called "ectopic" because they may arise in one or more unpredictable sites along the axon. You may recall that in a normal nerve the ion channel proteins are strictly regulated and most often consolidated at Nodes of Ranvier in myelinated axons. A damaged or recovering nerve may have too many or too few of its expected ion channels, or it may have inserted a different type of ion channel protein in its membrane. The result may be more excitability with pain or less excitability with loss of its transmitting role. Numbness, a "negative" sensory symptom results if there is a failure of transmission! Many of the axon proteins, although not all, are synthesized in the cell body in a distant ganglia. To reach the ends of the nerve, proteins are transported down the axon by anterograde axoplasmic transport. The cell body senses an injury to its branches and changes its personality in response altering its transcription of mRNA and translation of proteins, some newly expressed. Neurons thereby change from a stable unflappable transmitting status, to a repair or regenerative type. The repertoire, placement and interactions of new proteins, including ion channels renders electrical behavior that is now different than normal.

The local microenvironment of axons within nerves also changes after injury. Local chemicals, such as acidic hydrogen ion from lactic acid and others encourage axons to adopt hyperexitable behavior. Bradykinin is a small inflammatory pain peptide containing 9 amino acids that is found in injured tissues and activates specific receptors found on nerves. Histamine is a small but widely acting molecule that also interacts with several receptors and causes burning pain or itching by activating nearby axons. Many additional inflammatory mediators, released by macrophages and leukocytes, participate in generating pain and their numbers are growing. Further examples include ATP, substances released by inflammatory cells called cytokines and chemokines, prostaglandins, proteases (enzymes that degrade proteins), serotonin, norepinephrine and others. We will meet an interesting pain mediator also locally released at sites of injury, nerve growth factor (NGF), when we discuss regeneration below. Recombinant human NGF (rhNGF) was administered by injection during several clinical trials mostly unsuccessful, over two decades ago.[99] However during the Phase II trial, rhNGF was noted to generate deep aching muscle pain where the injection took place. One unfortunate individual experienced an overdose of NGF and developed a widespread problem of aching muscles!

Neuropeptides also have interesting roles following an injury. We have already discussed SP and CGRP. After nerve damage neuropeptides are released and accumulate within the injury site, causing, among other actions inflammation directly within the nerve trunk. This involves rises in local blood flow, "plasma extravasation" with swelling and pain generation. Both neuropeptides SP and CGRP activate mast cells within nerves. Mast cells in the airways and nasal passages cause allergic reactions by releasing the content of their granules containing histamine and serotonin. Histamine, in turn, generates both itching and pain. Thus we have come full circle! Inflamed and injured nerves release peptides that cause warmth, redness and swelling and also activate mast cells to release histamine, a pain signal! Overall the pattern is that of "neurogenic inflammation," similar to that described of other tissues, but that now targets the nerve trunk itself. How is this possible?

The outer sheaths or epineurium of nerve trunks have their own "nerve supply" with self-innervating branches called nervi nervorum. Nervi nervorum, perhaps thought of as freeway exits, arise from parent axons travelling down through the nerve trunk. Periodically within the nerve, a small sensory branch exits the nerve, leaves the endoneurium and meanders through the epineurium of the nerve trunk. While these small branches have been recognized for some time, understanding what they do has been challenging. Over 35 years ago, Drs. Arthur Asbury and Howard Fields suggested that nervi nervorum could contribute to neuropathic pain that develops during inflammation in neuropathies.[100] Disorders with prominent pain and inflammation of the nerve, such as Guillain Barré syndrome and vasculitis might activate them. Indeed physiological studies have recorded impulses from activated nervi nervorum[101] and in our own work, we noted that SP and CGRP were released by nervi nervorum after injury.[102] In other words, auto-neurogenic inflammation occurs, not unlike self-flagellation. Could this also explain why damaged nerves are so painful? Possibly, but it is not the whole story.

Hyperexcitable axons can set off different kinds of inappropriate signals. There may be random AP discharges, bursts of discharges or constantly firing discharges. We know that the number of discharges that occur for a given axon, the numbers of axons that are misbehaving and the frequency, or firing rate of the discharges all influence how pain may be felt. Dr. Marshall Devor and colleagues have shown that specific drugs can shut these discharges off and eliminate pain.[103] Anti-epileptic drugs are an important group of medications that curtail epilepsy by shutting down the inappropriate electrical discharges of the brain that cause seizures. They specifically impact ion channels. These drugs also silence inappropriate APs in discharging nerves by similar means. However the difficult news is that they do not offer complete pain relief. To understand why, we have to work our way up the nervous system to the next important part, the ganglia.

You will recall that ganglia house the main sensory neurons that send branches out into the body, and other branches that connect to the spinal cord. Ganglia have a major role in pain firstly because they synthesize most of the ion channel proteins that generate pain. The second reason is less obvious. Like axons and all other cells, ganglion sensory neuron perikarya have membranes composed of a lipid bilayer (two layers) with inserted proteins, including ion channels. Like their axons, perikarya are electrically active despite the fact that they are not aligned within the direct pain transmission pathway. As mentioned earlier, they are off ramp stations, like truck stops along the peripheral nerve highway. However, after an injury or inflammation of their axons this noninvolvement changes. Ganglion neurons begin to discharge on their own, generating spontaneous "ectopic" APs similar to those of abnormal axons. Some have likened their actions to pacemaker cells in the heart that normally send out cardiac impulses. "Pacemaker" cells generate regular repeating discharges to maintain regular heart rhythms. When they fail, cardiologists put in an electronic version, a pacemaker, never as elegant as the biological type. Devor and colleagues showed that not only do abnormal discharges of injured nerves arise from the axons and their branches but they also arise from ganglion cells. Thus after a nerve injury, one might think that freezing or silencing the axons with a drug locally at the site of injury might relieve the pain. However, ganglia neurons now take up the "injury flag" and send off discharges of their own. This is not a very helpful type of pacemaker. This is also another reason why neuropathic pain can be so difficult to treat.

The complexities of the story continue. The dorsal horn of the spinal cord houses a wealth of complex neural machinery that filters and controls pain. Disruption of the dorsal horn itself can cause neuropathic pain, as discussed. However, we also know that a nerve or axon injury, far out in the limb and nowhere near the dorsal horn causes horn circuitry to remodel in unhelpful ways. Some investigators suggest that the likelihood of remodelling depends on the intensity and duration of the discharges that have come its way. What kind of remodelling? This may include recruitment of nerve pathways and circuits that do not normally signal pain. For example, the neuropeptide SP is usually found in "normal" small caliber pain axons and neurons. In a chronic pain syndrome, SP may now appear in other neurons, for example in those that usually transmit touch sensations. Is this why a normal touch now becomes painful, the problem we described as allodynia? Pain sensations are usually confined to specific parts or layers of the dorsal horn spinal cord grey matter. In chronic pain, other parts of the dorsal horn now participate, enlarging and changing the pain circuitry. In the normal gray matter of the spinal cord, neurons, their connections and their branches are specifically layered in what are known as laminae (Fig. 33). Each lamina has unique interneurons and connection circuitry. Spinal cord laminae were first described by Bror Rexed (1914–2002), a Swedish neuroscientist. Bror Rexed is also famous for having made it easier for Swedish people to address one another informally [https://en.wikipedia.org/wiki/Bror_Rexed]. Surely this is almost as important as layers of the spinal cord, at least in Sweden!

The best way to think of sensory Rexed laminae is to consider layers of an onion. On the outside or the dorsal, or back horn of the gray matter are Layers 1, then Layer 2 is the next inner layer etc. Of course the gray matter of the spinal cord is not round like an onion, so this teaching aid breaks

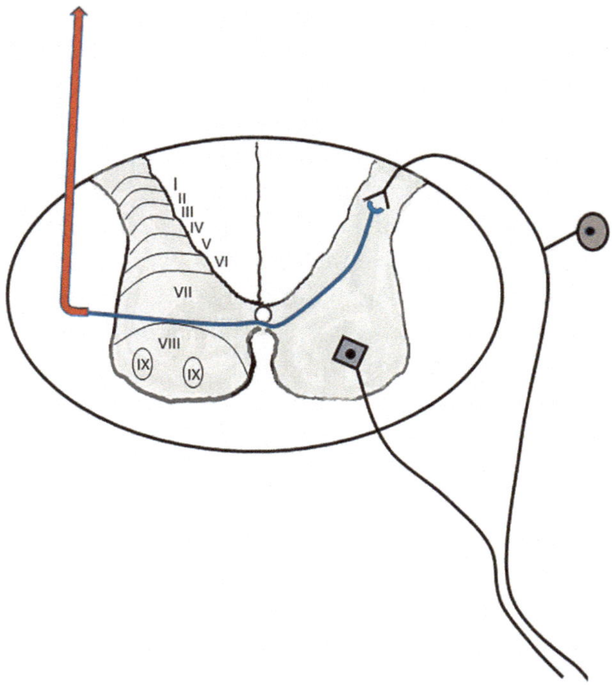


**Fig. 33** Simplified anatomy of the dorsal horn of the spinal cord showing the layered Rexed laminae and the synapse connecting an afferent sensory branch to a projection neuron that travels (blue) in the contralateral spinothalamic tract (red arrow) to the thalamus of the brain. This is a critical pathway for fibers carrying sensation for pain and temperature.

down when we get to the ventral or front part of the spinal cord, which has its own anterior, or ventral horn. Rexed layers house their own types of neurons. All of this is laid out exquisitely, reminding us how exacting the development of the nervous system is. From a single cell, we eventually arrive at layers each containing hundreds to thousands of specialized neurons. From the ganglia, the "central" branches of sensory neurons enter the spinal cord and end up in a Rexed layer. If they are small unmyelinated or small myelinated branches that transmit pain information, they arrive in layers 1, 2 and 5.

While controversial, some investigations have suggested that branches of pain neurons may travel to the incorrect layer after their axons are injured out in the limb. How does the brain interpret this strange wiring? One possibility is that the pain becomes associated with layers that are responsible for another kind of sensation, like light touch. Abnormal branching of pain pathways into "wrong" layers of the spinal cord has been yet another explanation for allodynia. However, branching, if it truly exists, is not the complete story. Like injured axons or DRG neurons, CNS neurons within the pain pathway can also become hyperexcitable. These neurons may be more likely to send inappropriate pain signals with little provocation. Neuronal genes are very sensitive to all sorts of events, from injury to inflammation to simple activation and like peripheral neurons, pain pathway CNS neurons can be persuaded to produce new proteins and to change how they behave. Overall then there may be several reasons for neurons to misbehave as neuropathic pain develops. They may send branches or sprouts to areas where they normally do not belong. They may have a change in their character, more likely to "fly off the handle." Another change involve other cells they interact with in the spinal cord.

Recall that the CNS, like peripheral nerves, has several cell types within it, not simply neurons. "Glial" cells include astrocytes, oligodendrocytes and an interesting cell type known as the microglial cell. Astrocytes and oligodendrocytes take on the roles of the SCs in the CNS offering support and myelination respectively. Microglial cells are not found in peripheral nerves and are the inflammatory cells of the CNS. They multiply, move and interact with all of the other cell types. Whether they benefit the CNS or not is debatable. For example, microglial cells rapidly multiply and form zones of inflammation in MS or encephalitis (brain inflammation from a virus or other cause). What was not recognized until recently was that injured peripheral nerves out in a limb, trigger a microglial response back in the CNS. For example, on the side of the nerve injury there ensues a vigorous microglial cell activation in the dorsal horn of the spinal cord. The microglial cells begin to cluster, somehow sensing that something is awry out there in the distant nerve.

Michael Salter and colleagues at the University of Toronto discovered something remarkable about clusters of microglial cells that form after injury. They found that they were not simply curious bystanders but became closely involved in how neuropathic pain develops and where. By way of their nearby presence, microglia were found to send molecular signals to neurons, altering their behavior and character. Microglial cells might be thought of as major irritants to the system because of their capacity to synthesize an army of molecules that are algesic, or pain producing. Some of this activity likely also damages neurons that normally filter and dampen pain or alternatively encourages pain generating neurons to send out new and inappropriate sprouts. These new findings have changed the conversation by making the point that pain is not always about the neuron! Along these lines, research is examining whether a blanket silencing of troublesome microglial cells may help to relieve pain. One common rule in biology however is that what you might gain by such a blunt approach is likely matched by a new problem. Microglial cells exist for a reason and suppressing them may bring on unexpected responses. Perhaps a better strategy would involve selectively interrupting only those signals between microglia and neurons that are irritating, so to speak. In this way new therapeutics may emerge out of the complexity of the interactions.

And yet the story goes on. Pain pathways travel from the dorsal horn, up the spinal cord to the brainstem where activation alerts us to pain. Within the brainstem there is a cascade of circuits known as the "ascending reticular activating system," or ARAS. ARAS is thought to support wakefulness and arousssal; it is disrupted in coma. In contrast, pain pathways arouse us and get our attention by connecting to the ARAS. With the ARAS, the next relay for pain pathways is the thalamus introduced earlier, a clearing station for almost all forms of sensation, including pain. Like the dorsal horn, thalamic alterations induced by chronic neuropathic pain involve branching of axons to make inappropriate connections, irritation by proliferation of microglial cells and the fostering of hyperexcitable neurons.

The thalamus next sends fibers upward to the cortex, the final pain destination. By this time our sensory signal has made connections with the ARAS, passaged through the thalamus, and entered the cortex of the brain. It is now safe to say that pain has gotten our attention. In fact some might suggest that the term "pain" is not meaningful until we include its impact on the brain. Unless we are made aware of pain, is it really pain? Researchers have linked selected parts of the cortex to pain using sophisticated imaging techniques such as functional MRI. Functional MRI combines the beauty of an anatomical image of the brain with its activity highlighted using colored (the computer program applies the color) patches or maps that "light up" when activated. The signal comes from oxygen use by active neurons, called a "BOLD" (blood-oxygen-level dependent) signal. For example if you move your arm, the motor cortex on the opposite side of the brain "lights up" from the BOLD signal. Pain generates a unique BOLD map or signature, lighting up sensory portions of the cortex. This includes the area posterior to the central gyrus known as the somatosensory cortex in addition to other parts of the brain including the thalamus. However, since researchers believe that parts of the brain cortex structure might be altered by chronic neuropathic pain, the pattern of MRI activation may differ. Some areas might enlarge or occupy parts of the cortex not usually involved while other areas of the brain are thought to shrink in size! The cingulate gyri, "hills" of the highly folded brain cortex that face inward toward the midline, are among the best understood parts of the cortex linked to chronic pain. By the time pain sensations reach the cortex, they have been extensively filtered and parsed by the spinal cord, the thalamus and other areas. They are no longer "raw" signals but are now part of our consciousness. As a result, they are heavily dependent on what else is going on in your cortex. If you are paying attention to something else, the pain may seem less intense. However, pain may feel more troublesome if you are anxious or depressed. Experiencing pain day and night interrupts sleep and promotes depression and anxiety, a vicious cycle! The cingulate gyrus may influence how much we care about pain: whether we are apathetic about it or consumed by it. Our relationship with pain becomes immeasurably more complicated as it interacts with the full repertoire of ongoing cortical activity and consciousness.

In summary, our inappropriate or ectopic discharges sent upward along the nervous system by damaged peripheral axons have quite a journey. They are filtered, altered, dampened or magnified en route. The pathways traverse the spinal cord, brainstem, thalamus and finally the cortex. They interact with other neurons and with microglia and at each step, the final pain experience can change. If chronic, the pathway changes may be long lasting. One can only appreciate how powerful this pathway is.

Pain pathways are "evolutionarily conserved" meaning that versions exist in most organisms that have a nervous system. Pain is designed to protect us, but when the signals go awry, perhaps from neurological damage, chronic neuropathic pain results. You may also appreciate how this complexity makes it very difficult to treat. A few words on treatment are needed.

Pain pathways are redundant. This means that there are several routes available through the nervous system to provide this information to the cortex. It also means that deliberate neurosurgical interruption of a single pathway is unlikely to alleviate the problem. Some suffering from chronic pain may request "Doctor please just cut my leg off-I can no longer live with this pain." Experiences with ablative surgery for pain are mixed nor have been rigorously evaluated in clinical trials. It is an unfortunate curiosity and outcome that amputating a limb with chronic pain does not eliminate the pain. Amputation creates a new cohort of injured nerves at the surgical site while all of the circuitry remodeling we have discussed remains in play, from dorsal horn to cortex.

In contrast to surgical treatment of pain, researchers have designed clinical trials to test oral medications. In theory several parts of the pain pathway might be targeted in this way. These agents and trials have offered mixed results however, despite the billions of dollars spent on them. The best known pain treatments are opioids, based on the opium poppy seed. The large family of opioid drugs include both synthetic types (opioids) and natural versions (opiates) from morphine, to heroin, methadone, codeine and oxycodone. How do they work? Opioid drugs target several types of opioid receptors on neurons and axons. The mu opioid receptor, from the Greek symbol μ is the best known. While mu opioid receptors are found all along axons, their classical locale is on afferents entering the dorsal horn of the spinal cord. At this juncture, mu opioid receptor activation inhibits incoming traffic from sensory afferent axons. Since mu receptors are in several areas of the nervous system however, they have serious and unwanted side effects such as addiction, drowsiness, respiratory slowing and constipation. Recall that pain pathways are closely connected to the ARAS, the pathway that is important for alertness. Activation of mu opioid receptors suppresses the ARAS. Drowsiness may be so severe as to slow and stop breathing, the reason that opioid overdoses cause death.

Clinical trials show that opioids help to relieve neuropathic pain. Like most pain therapies however, the relief that opoids provide is only partial or incomplete. Escalating doses are taken in an attempt to relieve pain, magnifying the risks of serious side effects. Before the magnitude of the opioid crisis was fully appreciated, guidelines supported their limited use in patients with severe nonmalignant neuropathic pain when side effects from other treatments left few choices. In persons with malignant cancers who are not likely to survive their disease, using opioids in careful doses has been compassionate. In patients with non-malignant pain however, recommendations have changed and guidelines now recommend largely avoiding them in favor of other therapies.

Currently, recommended neuropathic pain treatments, based on clinical trial evidence, involve anti-epileptic drugs, anti-depressants and gabapentanoids. None are "magic bullets" and their impacts are partial. Anti-epileptic drugs such as carbamazepine and phenytoin suppress unwanted ectopic discharges in hyperexcitable neurons. This requires blocking and limiting the function of ion channels in the axon or neuron. While anti-depressant medications help the depression that accompanies chronic pain, their activity depends on other actions. They are known as selective serotonin reuptake inhibitors, or SSRIs because they help to maintain higher serotonin levels by preventing its reuptake by neurons. They slow its clearance from the site of action. Having more serotonin available, in turn, dampens pain circuits, especially in the dorsal horn of the spinal cord. Some SSRIs also are called SNRIs because they also increase levels of NE, our autonomic neurotransmitter, that also offers actions to help suppress pain and relieve depression. Examples of SSRIs or SNRIs for neuropathic pain include duloxetine, amitriptyline, nortriptyline and venlafaxine. The final group of neuropathic pain treatments are the gabapentanoids, named after gabapentin, the first member of this group. Originally designed as an anti-epileptic agent, gabapentin is thought to act by inhibiting a specific part of a calcium ion channel known as the alpha-2 delta ($\alpha 2\sigma$) subunit of the voltage dependent calcium channel. Calcium ion channels of several subtypes play a very important role in axon and neuron hyperexcitability, arguably as important as the sodium channel. While not discussed in depth here, selective interruption of calcium signaling, beyond that offered by alpha-2 delta ($\alpha 2\sigma$) inhibition, are among new avenues being considered for suppressing neuropathic pain. Pregabalin is a newer member of the gabapentanoid family. While gabapentin and pregabalin are nonaddictive and do not cause death, they are not free of side effects and do not offer complete pain protection. Some have recently argued that they are overused and somewhat addictive although not comparable with the highly addictive opioids. Side effects of gabapentin and pregabalin at higher doses include dizziness, declines in memory, weight gain, swelling and fatigue. Cannabanoids, that include the active ingredients from cannabis are widely used in pain by persons with neurological disorders but their exact role is yet to be established. Tetrahydrocannabinol (THC) has cognitive side effects that complicate its use in pain. Cannabidiol (CBD) may offer more promise for pain but detailed trials are in the offing. There are, of course numerous additional cannabanoids and most have varying avidity for the Cannabanoid receptors, named Type 1 (mainly brain) and Type 2 (especially immune cells). Cannabanoid receptors are GPCRs, as discussed earlier under "Taste."

Given all of these imperfect options to date, it is not surprising that there are large numbers of additional bogus "cures" for desperate persons. I recently drove past a shop in a major Canadian city with its brightly lit sign stating "I cure pain." I don't think so. Very few "alternative" treatments have had rigorous clinical trials. Sometimes the term "alternative" itself is a problem suggesting an "alternative" to careful, rigorous testing for efficacy and safety. Untested approaches may not work, may be expensive and are not without side effects! More genuine, carefully developed candidate approaches to treat neuropathic pain are required. Basic, or discovery neurosciences research on pain mechanisms is essential toward finding these new approaches.

The final issue about chronic pain worth discussing is Chronic Regional Pain Syndrome (CRPS). CRPS is a striking disorder, easy to recognize but difficult to understand or to treat. In essence it is a chronic pain syndrome in one limb. A curious aspect of CRPS is how its chronic pain is associated with other phenomena such as swelling and color change. Persons with CRPS may be reluctant to use their warm and red or cool and blue limbs because of the pain, seemingly paralyzed. However, muscles and movement are not directly damaged by CRPS. Reluctance of a CRPS person to move their limb has other consequences including muscle loss from disuse, "frozen" immobile joints and osteoporosis of underlying bones. CRPS has been subdivided into types I and II respectively without and with associated nerve damage. In type II CRPS, while there is mild, moderate or severe nerve damage, how this generates CRPS with limb swelling and change color is mysterious. It is even more baffling to understand it when the nerve has not been damaged at all!

CRPS was misnamed "reflex sympathetic dystrophy" or "RSD" for many years. Fortunately clinicians have recognized that CRPS is not necessarily a reflex, may not involve the sympathetic nervous system and is not a muscle dystrophy![104] Sympathetic nerves and their receptors have close connections with sensory nerves, perhaps explaining their relationship to pain. The RSD label arose because blockers of sympathetic nerves seemed to relieve the pain when administered into the arm. Some instances of surgically cutting these nerves, a procedure known as sympathectomy, appeared to help. However, the numbers and quality of clinical trials for these treatments were limited. How activation of sympathetic nerves caused swelling or redness was also unexplained. Recall that peptides like CGRP and SP released from nerves cause warmth and swelling through vasodilation and plasma extravasation. In contrast, sympathetic axons constrict blood vessels causing skin pallor without plasma extravasation. One might surmise that the sympathetic axons somehow signal pain fibers in the limb to become activated and to release peptides. However, Dr. Jose Ochoa, a neurologist in Seattle suggested that RSD really did not exist as such but might be renamed instead as the ABC syndrome![105] ABC stands for "angry backfiring nociceptors," a rather accurate description of what CRPS may entail! Peptide containing unmyelinated axons, if retrogradely activated might be convinced to locally release neuropeptides, a process we previously described in neurogenic inflammation (Fig. 20). Their actions, in turn, might account for the tissue changes of CRPS. As in neurogenic inflammation, retrograde activation means that an AP signal moves antidromically, or the reverse direction of how sensory signals are normally transmitted, instead into the skin of the CRPS limb. Local delivery of active peptides ensues, released by "angry" or irritated ABC peptidergic nociceptive axons.

In support of CRPSII arising from ABC axons has been the rare finding that limbs with nerve injuries, have redness and swelling that is strictly confined to the injured nerve territory.[106] These persons have a pattern of skin changes involving some fingers or parts of fingers that correspond to the nerve supply. In full blown CRPS, what might activate ABCs in the limb is not really understood. Why intact nerves participate is unknown. Needless to say CRPS is a difficult condition to treat, explaining the heroic attempts at surgically ablating sympathetic nerves! Sufferers become desperate for any approach that might help alleviate their pain. Anatomical disruption does not accomplish the task. Physiotherapy and remobilizing a CRPS limb are important in recovery.

# CHAPTER 8

# Hope and change: Regrowth of nerves

Traditional ideas about the nervous system are based on the premise that it is "hard wired," like the electrical plan for your house. This concept describes neural pathways as elegantly and deliberately structured for key connections. Once functioning, more growth and sprouting is considered unnecessary. Neurological disease in turn is thought to interrupt and destabilize these precise patterns and circuitry. For example, unchecked growth of neurons might lead to tumors such as neuroblastomas from immature neurons or neuromas from misdirected sprouting axons.

## The adult nervous system is remodelling

The idea that ongoing growth and remodeling of the undamaged nervous system might exist is relatively new. However, when we consider its place in the body, a static and unchanging nervous system may not serve the best interests of an animal or human. In this chapter, we discuss not only how regeneration of nerves can occur, but also the case for growth as a normal state of affairs.

The most obvious case for ongoing axon growth is within the skin. The epidermis and dermis of the skin are infiltrated by large numbers of axons that provide us with normal sensation. However, we are also aware that skin cells, called keratinocytes, are normally shed, sometimes in the form of dandruff and house dust! The skin is under a constant state of remodeling, losing outer layers as the cells die and replacing them with new cells from deeper layers. Since skin axons travel into the epidermis of the skin they interact with a changing population of skin cells that surround them. The recognition that skin axons continue to grow within this changing environment is recent. Skin axons adapt well to "change management," unlike the human workforce! The fine branches that enter the epidermis take turns

*Our Wired Nerves*
https://doi.org/10.1016/B978-0-12-821487-9.00008-8

and twists as they navigate among the keratinocytes. A fascinating question is what factors determine precisely how many axons should innervate a given area of skin.

## First clean the mess

We learned earlier that damage to axons, called axonal degeneration (AxD) is irreversible. Once set in motion, there is no turning back. All parts of the axon from the injury site outward begin to dissolve and degrade during AxD. We also learned that an axon, once separated from its neuron parent cell body, rapidly undergoes AxD.

It was once thought that axons detached from their connection to the cell body degenerate simply because of a lack of support. Recall that proteins and mRNAs from neurons are shipped down the axon using axoplasmic transport. This includes ongoing delivery of mitochondria, essential sources of energy to keep it alive. Their transport is only part of the AxD story. Once the axon is cut or damaged, an immediate rush of calcium into the axon triggers active molecular destruction programs. Final steps include molecules called caspases and calpains that degrade axon proteins. Inflammatory cells make important contributions to AxD. These comprise white blood cells, especially neutrophils and macrophage/monocytes that are delivered to damaged nerves by the bloodstream.

AxD must be largely completed before regeneration is possible. Perhaps to be difficult, AxD leaves a trail of debris and inflammation. While some of this debris encourages growth, other components block or inhibit new growth. For example, degenerating myelinated axons deposit inhibitory molecules that signal new axons to back away. Several arise from proteins within the myelin sheath that originally surrounded the axons. Nogo-A was the first discovered of these and includes not just one, but two inhibitory portions respectively called Nogo-66 and Nogo d-20.[107] These names are somewhat amusing because "Nogo" simply means that axons would not "go" or grow when they encountered these inhibitory proteins. Nogo-66 is part of the original longer Nogo-A molecule and it has 66 amino acids (recall that amino acids are building blocks of proteins). Nogo d-20 is a more mysterious subportion of the protein that is also unpopular with axons. Composed of two unpleasant parts, Nogo-A assumes the title of a repulsive signal! Two other myelin associated proteins that inhibit growth are called MAG, myelin associated glycoprotein and OMgp, oligodendrocyte myelin protein. Nogo-A and MAG are found in both CNS and peripheral axon myelin whereas OMgp is found in the CNS myelin formed by oligodendrocytes. Given this additional inhibitor, CNS myelin inhibits axon growth more vigorously than peripheral nerve myelin. MAG is also somewhat famous in the clinical neuropathy world because persons can develop an autoantibody response to it. By attacking MAG, the immune system disrupts myelin structure and causes demyelination.

## Take my hand: Growth cones that lead the axon

Neurons navigate their axons through tissues using structures called growth cones.[108–110] Growth cones are extensions that emerge out of the ends of growing axons, also thought of as the "lead" portion of the growing axon (Fig. 34). They are sometimes described as battering rams, forcing their way through tissues but they also provide a helping hand, akin to constantly moving and exploring fingers. All growing axons require a growth cone to lead them toward a target and they are essential during neural development. Growth cones are also essential during regeneration of adult axons. However, in the CNS they do not get very far in their travels whereas peripheral axons have more success moving forward. Despite these advantages the success of peripheral axon regeneration remains substantially below expectations following nerve injury.

Growth cones resemble a hand or perhaps the webbed foot called a "palmate" found in ducks. The finger like extensions of the growth cone are called filopodia and between them are web-like structures called lamellipodia. Growth cones are constantly in motion, extending or retracting

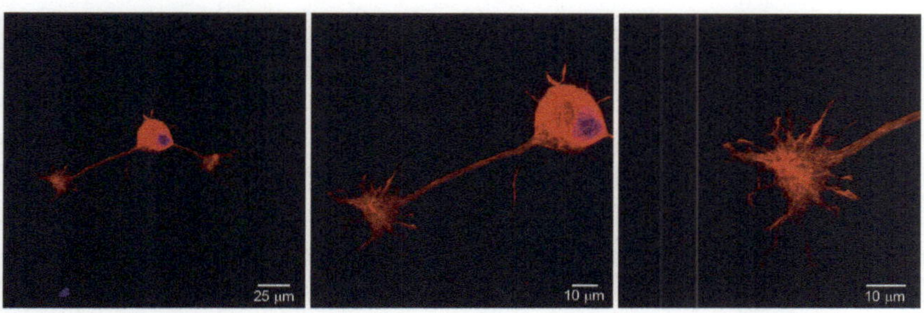


**Fig. 34** Images of an adult rat sensory neuron extending neurites (axons) with growth cones. The left panel is lower power illustrating two neurite extensions each with a growth cone. The middle panel is an enlargement of the larger neurite and the right panel is an enlargement of the growth cone. The neuron is harvested and cultured in vitro and the staining is for neurofilament (red) and DAPI (blue). *(Images courtesy of Dr. Anand Krishnan.)*

their filopodia while ruffling their lamellipodia and sensing what lies ahead. Once content with their landing site, the rest of the growth cone follows the lead filopodia, which then moves to the next target. Growth cones anchor themselves on a favorable spot, allowing the rest of the growth cone and the trailing axon to catch up. The base, or wrist of the growth cone is in transition, remodelling toward forming a new portion of the axon. To continue moving, a whole ensemble of proteins must work together. Firstly, sufficient amounts of cell "membrane" material are required to cover the growth cone and axon. Through the process of axoplasmic transport, lipid membrane material from vesicles must constantly replenish the growth cone. We learned about anterograde (forward) axoplasmic transport in a previous section.

How do filopodia work? A growth cone typically has several of them. As exploring fingers, they are constantly growing (advancing) or shortening (retracting). It makes sense that the advancing filopodia are seeking to drag the rest of the growth cone and the new axon along with it. One also has to imagine their navigation through its complex microscopic environment. They may meet other cells, proteins that are outside of these cells and all manner of connective tissue that lie in their space. Growth cones encounter basement membrane material, a mix of proteins such as laminin and fibronectin that often surround cells. For example, SCs lay down an extensive basement membrane that supports axon growth. On the other hand, if a filopodia has advanced into unfavorable territory, it shortens, retracts or may disappear altogether. A growth cone may elect to turn in a certain direction, a property known as "tropism." The filopodia elongate on the side they turn toward and shorten on the opposite side. Advancement, retraction and turning are all unique properties of growth cones and continue until they are satisfied they have reached the correct place. Following this lead, the remaining growth cone then follows suit. Filopodia are structured with a polymer protein called actin, arranged in long chains that shape the "finger." Actin chains or filaments, also called microfilaments, might be also compared to ropes or scaffolds. When a filopodia gets longer, new actin protein is added to the front end of the microfilament or growing "finger tip." At the base of the filopodia actin is dissassembed in the "hand" of the growth cone, and is recycled. An extensive series of molecules is required to add or subtract units of actin that include cofilins, profilins, Arp2/3, N-WASP, N-WAVE, VASP and others[42]; these require instructions, a signal to the growth cone from the environment to "come hither" or "go away." Specific receptors on the growth cone respond to extracellular signals and send instructions regarding actin assembly or disassembly and filopodia behavior. A specific extracellular cue may signal the filopodia to add actin and to grow. Alternatively, if the filopodia is not suitably placed, a repulsive signal may prompt actin subtraction and retraction. The absence of signals in a boring microenvironment or competing attractive and repulsive signals may direct a growth cone to remain in place. This is an oversimplification of a very coordinated operation.

Lamellipodia, resembling webs between the filopodia undergo "ruffling," a wavelike movement that helps direct forward movement. They are composed of a network of actin filaments but unlike the ropey bundles in filopodia, they form a meshwork, attached to the leading edge of its wave. Lamellipodia are also found in other cell types that migrate and move. They form the leading edges that fully anchor the growth cone, dragging the new axons along with them.

Are growth cones pushed forward, or do their filopodia fingers pull them along? The evidence, albeit controversial, suggests a pull. Growth cones also have an independent spirit and are capable of extension, retraction or turning even when their connecting axon has been severed! The success of their growth depends on something called "differential adhesion," or more simply, how sticky the surface is! Push from the contents of axon also occurs as a result of forward anterograde axoplasmic transport into growth cones. You will recall that microtubules are the "train tracks" that move vesicle packages along axons. Microtubules tracking from the axon enter into and terminate within the "hand" of the growth cone. Whether they unceremoniously dump their vesicular contents, or have a complex deposit process is unclear. However, as the growth cones move, you can imagine that the microtubules in the main body of growth cone are forced to remodel. They are unwound and unbundled then rebuilt and rebundled as new portions of the axon form.

What mechanisms allow the growth cone to be pulled and to move? Myosins are proteins that cause muscle fibers and other cells to contract and shorten. In growth cones they act to shorten scaffolds of actin filaments thereby giving filopodia and lamellopodia "contractile" properties. One might imagine rapid activation and deactivation of myosin in different parts of the growth cone hand associated with rapid and dynamic exploration of their environment.

Our descriptions thus far give some idea of the complex machinery at the front of a growing axon. How do they know where to grow? Correct pathfinding is essential to successful reconnection. Pathfinding depends on whether the growth cone sees attractive or repulsive cues. Think of yourself at a large gathering that you must traverse to get to the bar. On the left side of the crowd are people that are difficult to interact with or are unpleasant. At the party, on the right side of the crowd are attractive people you might prefer to meet and talk to. If you go left, the experience might be worth avoiding. If the crowd on that side is difficult enough they may even block your way. On the other hand, if you go right, the crowd is friendly and helpful. However if you plunge too deeply into them with good interactions and positive vibes, you may not make it to the bar. Everyone may want to talk to you, or perhaps shake your hand. What do you decide? Perhaps you might go down the middle, possibly the fastest way! You might just skim along the edge of the unpleasant crowd, avoiding them while getting pulled along by friendly faces, but not so many of them that you don't move or make it in time to the bar to order your drink!

During their travels, growth cones see a host of attractive and repulsive cues. Some may be only a little attractive and others a little repulsive. Moreover, many tissues lay down so called "gradients" where the amount of a cue gradually changes. On encountering this, the growth cone gradually gets the idea to slow, perhaps move one way or the other. If there is an attractive gradient, the growth cone may get irreversibly drawn into it and remain in place. Perhaps this is the target where the axon is meant to stop. Using our crowd analogy, think of yourself at the opposite end of the room from the bar. You deviate toward less unfriendly types and notice that this group seems to get friendlier and friendlier the more you head in their direction. Ahead of you is a yet more enjoyable prospect-the chance to order a glass of wine, with your best friends close by. You have reached your destination or target and are happy to simply stay there, among friends while you enjoy your Pinot noir.

Attractive and repulsive cue gradients form the basis for neurodevelopment. There are many examples, but perhaps the most striking involves spinal cord development. In newly forming spinal cords, axons are required to cross the midline to set up their proper anatomy. In doing so, first they encounter an attractive signal beckoning them to the other side of the spinal cord. However once they get there, repulsive cues now take over, instructing them not to come back!

While searching for the bar and for your glass of wine, its very difficult to see the precise location of the bar amidst the crowd of people. How do you know? To find the right direction, you need a guide. It can be a large fluorescent overhanging sign that says "Cash Bar"! Did you miss it when you first looked around? Or, the cues can be more subtle, like small signs with arrows arranged on the wall. A subtle cue might involve learning that three of the walls do not have the bar, so the crowd in front of the fourth wall must be at the correct target. Finally, you may have a friend (attractive, not unpleasant) who simply tells you the correct direction or points you to someone else, also friendly who is yet closer to where you want to be. This resembles what happens during neural development and perhaps also during nerve regeneration. The signs are like instructive molecules telling you to advance, retreat or turn. These could be small extracellular molecules perhaps on the surface of nearby cells, called guideposts. The friend who directs you is also a type of guide, resembling a guidepost cell along the pathway directing the growth cone. Repulsive cells may also be encountered, like unfriendly people that suggest going anywhere but near them!

The bottom line here is that regrowth and regeneration of axons initially depends on how the growth cone fares. If it constantly encounters hostile cells or molecules, it will not advance and the axon will not grow. If it identifies friendly guideposts and gradients of attractive cues, it will find itself going the right direction. What is the right direction for an axon that is cut, for example in the middle of the thigh? Does that axon know that it belongs perhaps to the third toe, or to muscle fiber number 37 of the hundreds of muscle fibers in the foot? Although there is debate, these detailed homing signals in adults probably do not exist. Motor axons are not likely to be labeled with their exact home address codes for specific muscles. Instead, they get hopelessly lost in the crowd. As a result, any new growth is a new adventure, and looking for guidance is essential, even in getting close to where they belong. In the case of neural development, a whole series of cues are set up to carefully steer axons in the right target. Without this more precise guidance, our developing nerves would arrive at random destinations without proper connections and without function. As the nerve grows during development first it might encounter a cellular guidepost, then a molecular gradient, and then be shunted over by a repulsive gradient. Along its journey, the axon itself might also change and begin to have new opinions over what is friendly or not. In this scenario, the maturing axon changes the proteins on its surface so that previously repulsive cues might now become attractive and vice versa. Beauty is in the eye of the beholder! At the same time, the tissue being navigated will also likely change. Have you ever tried to remember how to drive in a city you have not lived in for 30 years? Once you may have been looking for a movie theatre, now demolished. Now you might be more interested in finding an old friend's house!

Some axons may have specificity during regeneration in adults, namely in the optic and sympathetic nerves. This has been described as part of the "chemoaffinity hypothesis" in which some neurons remain permanently linked to eachother through a shared molecular interaction.[111–113] The hypothesis has been posed in the CNS but there is little evidence for it in regenerating peripheral nerves. Helpful guidance appears inadequate after nerve injury and is without the careful fidelity for example, to connect axons to selected individual muscle fibers.

Compared to neural development, the challenges during adult regeneration are much greater. There is generally no strict predetermined "pattern" that will guide axons to destinations. However some less refined cues or clues are present. For example many have wondered how motor axons might be generally directed to muscles and sensory axons to skin. Recall that most nerves are a mix of motor, sensory and autonomic nerve axons; to regenerate they cannot all arrive at the same destination! Some early guidance might be helpful in telling regrowing axons which freeway exit to use.

A serious case in point arises if the entire nerve trunk is transected and separated, as if the freeway is destroyed by an earthquake. Surgical reconnection may be required. If a severed nerve is resutured and reconnected by a surgeon, the axons from the proximal stump first traverse the site of repair then travel toward original targets. Using a microscope in the operating room, surgeons attempt to connect the epineurial tissue of both ends of the nerve, lining up the fascicles to match their previous orientation. For example, this may require suturing the upper or left part of the proximal nerve stump to the upper or left part of the distal nerve stump. While this may sound easy it is actually very difficult. Imagine a large nerve with many fascicles housing its axons. To surgically reconnect fascicle 4 on the bottom right, to its exact distal counterpart may not be possible even with a surgical microscope. Recall that nerves are normally under elastic tension. When they are cut, each stump shrinks down and moves away from eachother, leaving a gap between the stumps. It is technically difficult to re-unite and resuture them while orienting them as they once were. Sometimes a graft of tissue, such as a nerve from another site, or a biodegradable tube is used to connect these stumps.

Following repair, axons from the intact proximal stump, connected to their cell bodies, begin to sprout. The distal stump degenerates. Severed axons from the proximal stump experience a brief delay and adjustment to their new circumstances, as if "stunned" by their injury, and then consider advancing. They may send out many branches into the injury site. This dramatic branching from the stump can be very striking, as illustrated using genetically fluorescent axons, likely reflecting the neuron's attempt to plant at least one axon in the correct direction.[37] Once the axon sprouts find a satisfactory target resembling their original address, other axon branches with unsatisfactory placement can be eliminated. Witzel and Brushart found that a single parent axon, in the proximal stump may generate up to 40 sprouts! Excessive numbers of axon sprouts are a problem, in competition with one another for resources, called growth factors, discussed below. In neural development, competition for growth factors is considered to be a foundational principle that defines anatomy. A successful landing at a target allows the axons to soak up nearby growth factors. Unsuccessful sprouts wither away because they lack growth factors. This is called "pruning," as in a troublesome shrub in your garden and it conserves resources by eliminating unnecessary axon branches. The instruction for a branch to disappear is interesting and not fully understood. Pruning may share common mechanisms and molecules with AxD degeneration during neural development or adult axon injury.

One might imagine then that a series of competing axon branches emerging from the proximal stump of a transected nerve randomly choose sites to land on in the distal stump. Where they arrive may be completely spurious and unpredictable. Perhaps more land on sites of the distal stump that were aligned by the surgeon. Recall that in some extensive injuries, nerve trunks cannot be surgically reconnected, as in our famous Grizzly bear attack. In some instances, surgeons try to connect the gaps by providing a bridge for the axon branches to cross. Nerve bridges are made from nerve grafts from other nerves such as the sural nerve, or artificial implanted tubes to connect the stumps. Regrowth in these scenarios is yet more difficult than traversing a direct suture site and requires navigation across gaps, sometimes substantial. Think of this problem in the context of the English Channel, which for centuries warded off invasions from European nations. The presence of a gap or channel seems to dramatically add to the difficulties of an invasion.

While the destinations of axons may appear to be random, this is not always the case. Ideally one might want to discourage the wrong kinds of axons from reconnecting to inappropriate pathways in the distal stump. Random motor axons may travel into cutaneous branches headed to the skin. They will not be of much use there! Similarly, sensory axons may navigate nerves that connect to muscle fibers, also unhelpful. New axons do receive some cues directing them to appropriate major freeway exits.

Dr. Tom Brushart discovered "preferential motor reinnervation," or PMR[114,115] a process in which motor axons are more likely to choose a correct muscle branch and sensory axons choose cutaneous branches. PMR was demonstrated elegantly using fluorescent dyes to label axons and their journeys, but its success is incomplete. The challenge has been to improve the likelihood of correct targeting. For example, electrical stimulation of the proximal stump soon after an injury has been found to improve PMR, as discovered in further work by Brushart in collaboration with Dr. Tessa Gordon from the University of Alberta.[116] Successful PMR likely depends on molecular guideposts that new axons detect as they enter the distal stump of an injured nerve. Along these lines, motor and sensory axons prefer to growth through different kinds of basement membranes. Recall that basement membranes provide overall direction for growth cones to follow but are rudimentary direction signals without refined instructions for how to get axons to their exact targets.

Recall that after crush or injury of a nerve, the part of the nerve on the far side of the cut, the distal stump furthest away from the cell bodies undergoes AxD. If the crush or cut is complete, all of the axons in the distal stump degenerate. However, two important constituents remain, SCs and their basement membranes. SCs persist in the distal stump and once the nerve is injured they acquire a new role. SCs do not sit and grieve when their axon partners disappear but instead become active, migrate, proliferate and acquire a host of new proteins. Extensive basement membranes synthesized by SCs serve as essential guidance cues for new axon growth. After injury, SCs line up along the direction of their previous axon partners and their basement membranes form unique channels or pathways for new axons to grow along. If no partner comes their way for a long time, that is another story. The roadways designed to attract new axons are called "Bands of Bungner" after Otto von Büngner (1858–1905), a German surgeon (http://www.whonamedit.com/doctor.cfm/2471.html). Once an axon negotiates the gap or suture site of a nerve injury into the distal stump, there is help finding their way by Bands of Bungner. The distal stumps have subtle differences in their content of basement membrane molecules and growth factors that are cues for motor or sensory axons.[117,118] These have names such as HNK-1 (human natural killer-1)-a frightening name for a repurposed immune cell molecule found in nerves! Subtle differences in the molecular make up of growing axon tips influence the interactions of the new axons in their environment. NCAM (neural cell adhesion molecule) is a glycoprotein (proteins that have a carbohydrate portion called glycan) attached to axons that interacts with SCs and basement membrane. NCAM influences whether PMR is likely to occur. Further to this, Franz, Rafuse and colleagues from Dalhousie University in Halifax discovered that a specific portion of NCAM called polysialic acid, or PSA has a major influence on motor axon pathway finding.[119] PSA is thought to alter how firmly new motor axon sprouts adhere to basement membranes. If too firmly attached, they may be less amenable to the types of remodeling and pathway navigation essential toward forming reconnections to muscle. Thus, any new axon may have the ability to recognize its preferred local environment and basement membrane and to remodel accordingly, as in the case of a motor neuron that sends its axon to a muscle. Problems arise if a given neuron sends one branch to a relevant destination but errs by sending a second branch in a wrong direction. The second branch will likely be pruned but this concept is not yet confirmed.

How do axons act and function once they emerge from a proximal stump? When a nerve is cut, do the axons in the proximal stump all rush out together once they have regained their confidence? Do they invade the distal stump like a horde? Surprisingly, new sprouting axons are fairly circumspect and take their time. They test out the new environment using their growth cones. They may discover a Band of Bungner close by or alternatively encounter a gap not containing any helpful tissue. A surgeon may implant a tube connecting the distal and proximal stump. It will fill with blood constituents such as fibrin, followed by connective tissue cells like fibroblasts, and eventually blood vessels will form within it. While axons might next be expected to enter, migrating SCs precede them. SCs are like unruly children that axons need to chase down hallways. I previously described them as intimate partners of axons, but here we require a different analogy. When they are in contact with axons, SCs are on their best behavior and quietly remain in place. Stable compliant SCs associated with axons synthesize myelin and help to regulate ion channel placement within nodes, but appear to do little else. SCs aggregate smaller unmyelinated axons together in Remak bundles. However, if the parent axons vacate, perhaps from a more proximal axonotmesis injury, SCs become unruly. Their repertoire of proteins change. They begin to divide, migrate, and form elongated processes that guide later axons. The elongated unruly SCs, deprived of axon supervision, also synthesize basement membrane material we have discussed above, for attracting newly growing axons. If there is a gap, the SCs begin to migrate into the fibrin and among the fibroblasts that preceded them. They may explore this new territory closely with new endothelial cells destined to form blood vessels. However, SCs also like to migrate into uncharted areas at random, preferring to grow away from the orderly life in the proximal stump.

Although SCs lead the outgrowth and new axons follow, they do not follow readily. A single axon may venture out followed by a few axons then eventually by a crowd of axons. This important property, also highlighted by Gordon and Brushart, is called "staggered" regeneration or outgrowth.[120–122] The pattern indicates reluctant growth. If you were to section through a newly regrowing part of the nerve and examine its structure by EM, you might see "regeneration units" originally described in 1972 by Morris, Hudson and Weddell.[123–126] These authors characterized Type 1 units as housing a more mature myelinated axon clustered with a SC and several unmyelinated sprouts. A Type 2 unit was composed of a collection of unmyelinated sprouts only. While we might think of the Type 2 unit as the expected pattern from sprouting axons, the Type 1 pattern illustrates "staggered" regeneration with a single "senior" axon perhaps guiding several newer fiber sprouts. Within the newly forming nerve bridge, more mature myelinated fibers that have crossed the gap earlier, are now accompanied by newer axons, perhaps in clusters of sprouts.

In summary, some new axons strike out on their own as a lead axon, also called a "pioneer" axon. Others choose to follow. What makes a neuron decide to be bold or instead to hold back is unclear. Some axons prefer to grow together in bundles, probably with one or more long SC processes, and basement membrane material guiding them. The fine SC processes that intimately guide one or more sprouting axons may be difficult to discern and EM may be required. SCs processes may twist and turn and their partner axons closely follow. We have called this the "axon-SC" dance. Not only is there a close physical relationship but molecules are exchanged between the two. For its part, the SC makes the basement membrane that includes proteins called laminins or fibronectins. SCs also produce growth factors for nerves.

Laminins and fibronectins are complicated proteins. Laminins, of which there are at least 15 members, are shaped like a cross, with subunits that bind to molecules on cells and other subunits that allow it to polymerize, or weave together in long structures. This structure explains how they might bind axons or SCs to a surface. Fibronectins are long rod shaped proteins with repeating subunits that perform functions similar to those of laminin. During development, some sensory axons prefer specific subtypes of laminins and others prefer fibronectins. During regeneration, this may also be the case, with axons in the skin, for example, having a preference for specific laminin subtypes. Moreover, fibronectin levels rise in the area of injury during regeneration in order to help guide the movement of SCs and axons. Both axons and SCs have their own receptors that bind to laminin, fibronectin and collagen and are known as integrins. Like their binding partners, integrins have subunits and form several subtypes. To add to the complexity, these vary in the cells they appear on, and on what kind growth they support.[127–129] The diversity is breathtaking and spectacular.

Nerves in turn provide support for their SCs. At least two major signals for SCs arise from the axons. One, called Neuregulin also has several different subtypes with differing actions.[130,131] With this in mind, there is further complexity and diversity in this signal, briefly summarized here. The Neuregulin proteins belong to a family called epidermal growth factors, or EGFs. EGFs stimulate a variety of cells to proliferate, including the epidermal cells in skin. One form of Neuregulin, Neuregulin 1 (Nrg1) behaves like a growth factor with an important difference; it is attached or tethered to a cell membrane. This feature is unlike most of the soluble growth factors we will discuss below. Nrg1 subtype III is specifically found tethered to axons and positioned to stimulate SCs. Signaling by Nrg1 operates through ErbB receptors, part of a large receptor family called HERs (human epidermal growth factor receptor). HER receptors in turn, have their own notoriety for reasons that include their presence on breast cancer cells, a finding that helps to guide therapy. Meanwhile, tethered axon Nrg1 signals SCs to myelinate during neural development by stimulating both their ErbB2 and ErbB3 receptors.[132] During later stages of regeneration, remyelination of new axons is important for their maturation and resumption of functional connections. Nrg1 type III and Erb receptors may also support regenerative remyelination but this role is controversial. However, a somewhat different role has been considered. The amount of Nrg1 type III on a given axon increases as it expands in width, or radial diameter during either development or regeneration. Since axon radial growth is known to trigger its myelination, Nrg1 type III may act as a "size signal" to instruct SCs to initiate myelination. Finally, Nrg1 can be cleaved of its tether to the axon by "proteolytic" proteins (proteins that cut other proteins into parts) allowing it to float freely and attach to SC ErbB receptors. Beyond myelination, Nrgs signal a series of changes in the mRNAs, proteins and behavior of SCs. Other types of Nrg1 such as type I, autostimulate SCs, since they are synthesized by SCs but also act on themselves or neighbor SCs. The overall message is that signals like Neuregulins and their receptors are adopted by neurons and SCs from other cell types. With subtle alterations in their subtypes they accomplish an amazing array of actions.

A second signal from axons to SCs involves the neuropeptide CGRP we have discussed earlier. You may recall that small sensory axons contain CGRP and that they play a role in causing pain and inflammation. CGRP dilates local nerve blood vessels, vasa nervorum, that have an important role in the repair of damaged nerves.[133] However beyond these roles, CGRP also directly signals SCs to proliferate.[134]

There are probably many other examples of axon SC molecular interchange left to be discovered. The important points to remember are that "naked" axons are uncommon and that axons are largely inept without their SC partners. Some naked sprouts might traverse short distances, and there are instances in which an artificial environment can substitute for the SC to support more extensive lone axon growth. Surgeons and researchers have been thinking about this for a while, adding tubes with basement membrane molecules, growth factors and other attractions to coax axons to regrow into them. Success however, has not been overwhelming.

New questions have emerged. Why isn't the first wave of outgrowth more robust? Why not a "wave" of invading axons? In part this depends on the SCs moving ahead to lead the charge. They do so, but must weave their way through obstacles in the way. Axons are yet more hesitant, not only relying on SCs or pioneer axons to coax them, but generally less enthusiastic about early growth. This concept differs from common misconceptions arising for CNS focused work in which peripheral axon growth is viewed as robust without limits. CNS axon outgrowth is highly restricted but regeneration in peripheral axons is not exemplary, limited by exposure to hostile extracellular material. One of these unfriendly elements is called CSPG, which stands for chondroitin sulfate proteoglycan. CSPGs are the unpleasant people in the room described in our earlier analogy about cues. In fact, CSPGs are so hostile that in the laboratory they are often deliberately added as test molecules to assess axon growth rates, a kind of regenerative "stress test." If cultured neurons exhibit growth on a CSPG substrate after some form of manipulation, they are surely ready for any environment!

Growth cones have receptor proteins on their surface that signal them to stop growing or perhaps to retract. CSPGs signal advancing axon growth cones by activating complex "brakes" within them. These stop signals are of considerable importance in the field, since their existence opens possibilities that "releasing the brakes" could enhance regenerative growth and outcomes. Some brake molecules first present during neurodevelopment and are repurposed for adult neurons. One example is called semaphorin, a singularly unpleasant molecule first discovered in grasshoppers that blocks growth. Receptors on neurons called plexins can detect several types of semaphorins whereas CSPGs, in turn, interact with these molecules and their receptors to inhibit growth cones. Once a CSPG-sema-plexin signal is detected inside the growth cone, a whole series of additional molecules cause the growth cone to collapse and retract. One of the more important is called RhoA, a member of another family of proteins called GTPases. As the growth cone is exploring its environment, RhoA helps to brake or retract filopodia that encounter a repulsive cue such as CSPG. As you might expect, inhibitors of RhoA encourage growth.[135] The range and diversity of the molecular players that influence growth cone behavior is another example of detailed and exquisite peripheral neurobiology. I hope that the reader, not expected to memorize these many examples, simply gains an appreciation of the both the challenges and opportunities to intervene on behalf of patients with deficits from nerve disease.

Why are RhoA, Semas in axons or CSPGs involved with our regrowing axons at all? Why "brake" neurons when we want them to grow? Furthermore, why is a whole range of brakes built into a neuron? Recall that nerve regeneration is not a normal physiological process. Our bodies are not built or poised to respond to an upcoming nerve injury. In the absence of injury these molecules are probably designated alternative roles. They may prevent unwanted growth in order to facilitate a stable system of "wiring," our connectome. For example, if neurons in the brain, DRG or spinal cord were constantly in a state of growth and plasticity, one might wonder at what sort of neural anatomy would result! There might be connection chaos, an outcome that may be the case in certain genetic neurological conditions.

## What does snake venom and mouse salivation have to do with nerves?

In the past, the nervous system was thought to be "hard wired" and fixed. Adult nerves were viewed as static and incapable of significant growth. New concepts regarding the plasticity and dynamism of neurons have revolutionized neuroscience. Scientists have acquired exciting new tools to theoretically reverse neurological damage, setting the stage to translate discoveries into therapy. One of the major breakthroughs came with the discovery of nerve growth factor (NGF), the first of an expansive list of molecules that foster neuron growth.

NGF was discovered by Dr. Rita Levi-Montalcini (1909–2012), an Italian neuroscientist with a fascinating story [wikipedia.org/wiki/Rita_Levi-Montalcini]. As a female Jewish investigator living in fascist Italy before and during the second World War, her scientific work was crippled by discrimination and the war. However, she persevered, sometimes performing experiments in her home. After the war she moved to St. Louis in the United States to work in the laboratory of the pioneering neuroscientist Dr. Viktor Hamburger (1900–2001). In this laboratory, they made the remarkable discovery that sympathetic autonomic neurons grew extraordinary arbors of axons when exposed to extracts from sarcoma tumors of mice or salivary glands from mice or snakes.[136,137] Hence the reference to snake venom! Through collaboration with an American chemist, Stanley Cohen, who extracted the active ingredient, the group isolated the protein that caused this extraordinary growth and named it NGF. Based on their original discoveries in the 1950s, Rita Levi-Montalcini and Cohen won the 1986 Nobel prize. Some argue that Hamburger ought to have been included. Despite this issue, Levi-Montalcini became an indomitable icon of neuroscience and a role model for women in science until her death at the age of 103. The original images and drawings of NGF's action are beautiful to behold. One of the original techniques used by Levi-Montalcini and Hamburger was to grow sympathetic neurons as an "explant cuture." This is a simple preparation in which a whole ganglion is removed from a chick or rodent then is placed in a dish and fed nutrients to keep it alive. When NGF was added to the nutrient broth, a beautiful halo of outgrowing axons emerged from the explant resembling a sunburst. It must have been a truly remarkable experience to witness this dramatic event on the bench.

Not all types of neurons responded to NGF but its discovery led to the identification of other growth factors. NGF's impact was on sympathetic, small sensory and embryonic neurons. A series of protein family members related to NGF, each with its own peculiar impact on selected neurons has since been identified. Yves-Alain Barde and colleagues in 1982 extracted the first NGF protein relative from large volumes of pig brains, and called it brain-derived neurotrophic factor (BDNF). In contrast to NGF, BDNF was a growth factor for motor neurons or larger sensory neurons. Since its discovery, BDNF has been linked to many roles, too numerous to include here. Examples include impacts on pain signaling in the dorsal horn of the spinal cord and on memory formation in the hippocampus of the brain. Its importance has arguably overtaken that of NGF and the other growth factors. BDNF is a good example of a molecule with a major impact on one area of neuroscience, namely regeneration, but repurposed for a number of additional roles.

Overall, direct relatives of NGF are classified as members of the neurotrophin protein family. They include NGF, BDNF and two additional growth factors called NT3 and NT4/5 (neurotrophin 3, 4 and 5). Remarkably, separate investigator teams worked on NT4 and NT5 until it was realized they were working with the identical molecule! NT3 signals larger sensory neurons. NT4/5 acts on medium sensory neurons, sympathetic neurons, motor neurons and CNS neurons. To activate a neuron, each type of neurotrophin requires a unique receptor; hence neuron responsiveness depends on whether the relevant receptor is found on its surface. The classical receptor for NGF is called TrkA, which stands for tropomyosin receptor kinase A. Trk A receptors are found on small sensory pain signaling neurons and on sympathetic autonomic neurons. This explains why NGF has such an elegant, robust action on sympathetic neurons when first tested by Levi-Montalcini and Hamburger. TrkB receptors are activated by BDNF and are found on medium sensory neurons and motor neurons. TrkC receptors are found on larger sensory neurons that sense light touch, position and vibration. NT4/5 also binds to TrkB receptors, as is the case for BDNF. This alternative preference is called promiscuous binding for obvious reasons.

Since the discoveries of NGF and neurotrophin family members, a range of new growth factors with differing protein structure have emerged. As you might expect, they have their own unique sets of receptors on neurons to promote their action. For example, glial cell line-derived neurotrophic factor (GDNF), operates on some types of sensory neurons, motor neurons and on neurons from the CNS. It has two different receptors that must come together on the cell membrane to be activated. Yet another discovery was ciliary neurotrophic factor (CNTF), a protein from the ciliary ganglion, part of the autonomic nervous system of chicks. It is worth noticing that like NGF, BDNF, CNTF and several other growth factors for neurons were discovered from unforeseen directions. The lesson here is that unfettered and curiosity driven science pays big dividends and is more productive than targeted research. Work decreed to satisfy short sighted targeted priorities, such as rapid commercialization, is often unimaginative and frequently unproductive.

"Epidermal growth factors" (EGFs), "bone morphogenetic proteins" (BMPs), "platelet-derived growth factor" (PDGF), "pleiotrophin" (PTN), and "hepatocyte (liver cell) growth factor" (HGF) are yet other growth factor molecules. You will notice that they do not appear to have connections to the nervous system, instead targeting epidermis, bone and other tissues. PTN is "pleiotrophic" meaning that it has many targets. While all were identified from other tissues, these growth factors were later observed to facilitate growth in neurons. All have interesting nuances in how they act. For example, HGF appears to control growth in the sensory axons of the skin. When skin is shorn or shaven of hair, HGF levels in the skin rise and cause the nerves to sprout.[138] Why should this happen? This is the beauty of science. In exploring why such a strange series of events might happen, you often stumble across something further. By following the science, the paradox is that a commercial product may emerge more quickly than if proscribed.

As you might have learned so far, growth factors impact neurons widely or selectively and in several ways. Many are synthesized by SCs as part of their role in supporting neurons. By using a substrate signal (basement membrane) and a soluble signal (growth factor molecule) SCs have powerful and complementary tools to manipulate their partners.[127] However, as we have also emphasized, neurons also support and encourage other neurons. Some neurons themselves might make growth factors like BDNF and GDNF. An older maxim in neurosciences is that neurons that "fire together wire together." Growth factors support and facilitate these connections to work better and to be mutually beneficial. For example, why do sensory neurons contain both GDNF and HGF? Release of some growth factors by neurons allows them to act on their own receptors or those of their next door neighbor. This is called an "autocrine" action, a property we described earlier in SCs using Nrg1.

One of the more surprising neuronal growth factors is insulin. Produced in the pancreas, insulin regulates blood sugar levels, an important role that prevents the development of diabetes mellitus. As it turns out, insulin is also a potent neuron growth factor, an action unrelated to its impact on blood glucose. In an interesting paper in Science in 1972 called "Nerve Growth Factor and Insulin"[59] Frazier and colleagues emphasized that the structure of NGF and insulin are similar and that they may have evolved together. The authors correctly predicted that there may be shared roles. It came as no surprise that very low doses of insulin, given to neurons in a dish, encourages them send out robust axon branches.[139] Recall that the behavior of isolated neurons and their axons in culture, as described during the discovery of NGF, helps to predict their impact on overall nerve regeneration.

A technical explanation is in order here. When neurons are removed from a rat, mouse or other animal, and placed in a dish they grow and send out branches after a slight pause. This outgrowth depends on what growth factors they are exposed to. The branches that emerge out of these "harvested" neurons are a little different than if they were in the body. For example, a neuron sitting in the dish does not know which direction it should send its axons. The result is that they send out a beautiful arbor of branches in all directions, as we discussed with NGF (Fig. 35). The growth behavior of neurons, even adult ones "in vitro" can be striking to observe. The outgrowing branches are strictly called "neurites" by neuroscientists, although many equate them with axons. Recall that many neurons in the CNS have both axons to send out signals and smaller but complex dendrite branches to receive them. Thus, an outgrowing neurite process from a neuron in a dish could in theory become an axon or a dendrite. Sensory neurons are the exception in that they do not have dendrite processes as part of their structure. Fortunately we have also learned that when neurons in a dish ("in vitro") show exuberant growth in response to a growth factor, this often means the same growth factor will increase axon growth in the body ("in vivo"). Growth "in vitro" can help to predict whether a treatment approach might help nerves regenerate. Of course, this is not always the case. A new cancer drug may be highly lethal to cancer cells in a dish, but have no impact on a patient. A response in mice or rats does not guarantee an identical response in humans. Nevertheless it is a good starting

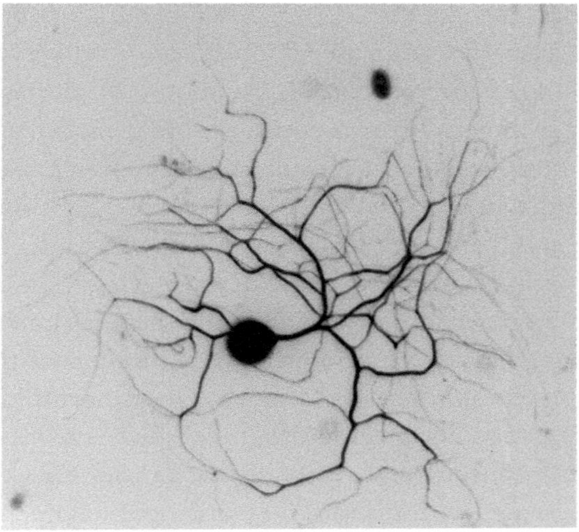

**Fig. 35** Image of an adult mouse sensory neuron harvested and cultured in vitro (in a dish) with extensive neurite branching.

point in the case of neurons and is preferable to the guess work that may characterize expensive human-only clinical trials. "In vitro" growth testing offers an important and more accurate screen to predict what molecules should undergo clinical trials.

Let us return to insulin. The exciting finding is that insulin not only has a major impact on growth of neurons in a dish, but it also improves regeneration "in vivo."[140] Insulin has receptors of its own (IRs, insulin receptors) as expected of any growth factor. IRs are found on most if not all neurons indicating that it acts widely on many types of neurons, unlike many growth factors we have discussed. Another interesting property of insulin is that its receptors are on the branches of axons, even in the skin.[141] We usually think of insulin as lowering our blood glucose. Diabetics have elevated glucose because they cannot produce insulin or because it does not work properly on its receptors in the body, particularly in liver, muscle and fat tissue. This latter problem is called insulin resistance. Type 1 diabetes mellitus arise from an immune attack on the beta cells of the pancreas that produce insulin, depleting insulin availability. In contrast, type 2 diabetes mellitus arises both from resistance to insulin and later from more gradual loss of beta cells. In an earlier chapter we discussed how diabetes damages peripheral axons in diabetic polyneuropathy. While much of the research in this condition has focused on how chronic high glucose levels damage nerves, a smaller group of investigators has suggested that lack of insulin or impaired insulin action could be important in the disease. How do you sort out whether it is the high glucose that damages nerves or if insufficient levels of insulin are to blame? If small doses of insulin are administered directly over nerves, their axons are encouraged to regenerate. If the mouse or rat has been diabetic, local insulin doses directed to nerves will reverse diabetic damage, but only in the specific nerve exposed to insulin.[142] Moreover, these low doses of insulin are insufficient to lower the blood glucose from its high diabetic levels. The response is an action that directly targets axons and neurons, not involving glucose levels. Diabetic nerve damage can also be reversed by injecting low insulin doses into the spinal fluid where it circulates and enters DRGs.[143] Finally, by making small injections of insulin into the foot of a diabetic mouse, sensory axons can be coaxed to reconnect to the skin.[141] The opposite foot that does not receive insulin does not recover. Thus, by using small local doses of insulin alone that do not impact blood glucose levels, you can improve regeneration and also reverse neuropathy in mice and rats. We do not yet know if a similar strategy also works in humans. Some of us are hoping to find out.

Inflammatory monocyte and macrophage cells are important sources of growth factors. Some macrophages enter the nerves and ganglia from the blood stream, while others simply live within nerves and ganglia, so called resident macrophages. Interestingly, their numbers grow after injury. The traditional view is that monocytes invade damaged nerves from the blood, where they are transformed to become macrophages. Unexpectedly, macrophage numbers also rise in DRGs housing the cell bodies of damaged axons despite their not sustaining any direct injury. These changes are remote, involving DRGs even though the nerve injury might be a meter away, down in the foot. Why ganglia react so empathetically and energetically to injuries of the distant axons is unclear. More unexpected twists to this story have emerged. Dr. Anand Krishan, a former postdoctoral fellow in our research laboratory (now on Faculty at the University of Saskatchewan) has discovered that many macrophages originate and multiply directly within ganglia and do not invade from the bloodstream.[144] These home-grown macrophages seem to collaborate closely with the perineuronal satellite cells, the intimate partners that closely surround neurons, described in earlier chapters. The architecture of the ganglia changes with new layering of macrophages and satellite cells around neurons. Why is this important? It may mean that some types of nerve and ganglion inflammation do not arise from the bloodstream, but involves local macrophages. The work highlights interesting cross talk among three separate cells: sensory neurons, resident macrophages, and perineuronal glial cells. Stem-like cells are also involved. Further work is required to understand the molecular conversation within DRGs and how it might influence regeneration and possibly the development of pain. This complex set of events may resemble an uprising or local riot of disgruntled workers within their place of work, the DRG, as opposed to a coordinated uprising by a national worker's party that invades from the bloodstream! Anti-immune therapies for neuropathies may need to target the correct group!

When inflammatory cells enter nerves and DRGS, they bring a series of inflammatory molecules with them. Low doses of some of these molecules might also promote growth! Too much of a good thing however can turn deadly. If inflammatory cells are too numerous or too "activated" they may spew out so many inflammatory molecules that neurons, axons and glial cells perish.

The mast cell is another cellular resident of nerves and ganglia. This unusual cell, introduced earlier when we discussed neurogenic inflammation, seems to have one main purpose in life. Its chief characteristic is that it houses large granules, repositories of molecules that are available for release. Their dumps are so large that fully "degranulated" mast cells are rendered invisible from view under the microscope. Mast cells are well known in the skin and lungs because they contribute to allergies. Some of their main ingredients are histamine that causes redness and itching and serotonin that impacts blood vessels, platelets and other organs. Excessive degranulation of mast cells in a person during an allergic reaction, known as anaphylaxis, can cause death. Mast cells are less common than macrophages and are randomly placed within the endoneurium of uninjured nerves. Their numbers rise after injury.[145,146] Mast cells also secrete NGF, BDNF and likely several other growth factors within nerve fascicles. It is not clear whether their release of growth factors is selective or requires catastrophic degranulation!

Why not simply inject growth factors into humans to cure neurological disease? This possibility has not been ignored. The discovery of NGF ignited thinking about its possible role in altering the nervous system. The application of growth factors however is a much more complex endeavor than imagined. We often do not know if a given growth factor is actually missing in a disorder or whether it is simply not effective in preventing or repairing disease. It is also possible that a growth factor might promote undesirable growth. For example, high doses of NGF had unusual side effects in animals. The sympathetic nerves sprouted dramatically and uncontrollably, into and around blood vessels where they did not belong!

In the late 1990s, clinical trials of recombinant human NGF (rhNGF) for neuropathies were undertaken. Like many clinical trials of a new drug or treatment, Phase I trials were first used to assess the safe use of NGF in normal people. Phase II trials examine safety and feasibility in target groups of people and sometimes include an early assessment of efficacy. However they are not designed to ascertain whether the agent truly has a benefit in patients. Phase III trials are carefully constructed larger randomized double blinded trials of the agent compared to a placebo, sometimes including varying doses of the proposed agent. A randomized double blinded controlled clinical trial requires that neither the physician nor the patient know whether they are receiving placebo or the agent of interest. After a prespecified period of treatment, outcomes are examined and the "code" is broken. Only at this time do we learn which persons received the proposed active treatment or placebo. Phase III trials are more expensive to conduct and frequently require more than one medical center to recruit sufficient numbers of patients. Hence they are called multicenter Phase III trials. Along these lines, rhNGF was synthesized for proposed Phase II and III clinical trials.

What happened was quite interesting. An unexpected property of injecting rhNGF into humans was local pain, a finding that has subsequently led to new ideas over how NGF interacts with sensory axons. The pain, developing soon after injection was deep, aching, and unlikely to arise from new growing axons. Its presence also allowed subjects to identify the difference between rhNGF and placebo that they had been injected with during the trial. In other words, the trials were difficult to "blind."

You might predict what happened with rhNGF trials. These trials were for HIV infection related neuropathies and for diabetic polyneuropathy. In diabetes, the Phase II trial was reported as positive, suggesting that diabetic nerves were deprived of NGF and that replenishing it might be a "magic bullet." However, the subjects who received rhNGF were later asked whether they could tell if they were randomly allocated to the active agent, rhNGF or to the control placebo. In most clinical trials there is a "placebo effect" in which subjects can feel better despite not actually receiving an active agent. Subjects were able to predict which treatment they had actually received. The Phase II trial of rhNGF looked promising but based on the patients' responses, it was unblinded, a serious technical flaw. This "unblinding" invalidated the conclusions of the trial. This is a cue to remind all why careful quality measures of research are critical. In the subsequent Phase III trial, the control treatment included a mild pain inducing substance in the mix. With this newer trial design, patients could not tell whether they had received rhNGF or placebo because both caused some pain after injection. However, the trial was negative, without a benefit from rhNGF. In addition to these difficulties, scientific questions around rhNGF arose during the trials. We know that NGF only acts on TrkA positive small sensory axons and we also know that diabetes targets a variety of axons in neuropathy. Some small axons that are damaged by diabetes are TrkA negative, and this "nonpeptidergic" population is unresponsive to NGF. Thus the idea that rhNGF might cure polyneuropathy may have been naïve. A second concern was that the trials did not measure actual axons in the skin expected to grow back in response to rhNGF. The dose of rhNGF may have been too low to show a benefit. Finally a Phase III trial of rhNGF for HIV neuropathy, that involves smaller TrkA axons, did show a slight benefit, albeit not a dramatic one. In retrospect, things could have been done differently. These expensive trials had consumed their budgets without impressive impacts. A more recent proposal to use anti-NGF agents to alleviate pain in osteoarthritis adds another interesting twist on this story. While giving extra rhNGF was not of benefit, removing it might not be advisable; there is a possibility that low persistent or maintenance levels may remain important to nerves.

Other growth factor trials, including BDNF, NT3 and others in both PNS and CNS disorders followed. It is probably safe to say that none were a "home run." Some trials have tested growth factors added into the conduits or tubes designed for connecting injured transected nerves. Many variations have been tried including embedding growth factors onto materials within the tubes. Again, the jury is still out. Yet another strategy has been to engineer cells that synthesize large amounts of growth factors and deposit them into areas of nerve injury. One potential difficulty with this kind of strategy is that growing axons may simply pause in areas of plenty. Recall the analogy of the crowded room with attractive and repulsive cues. When axons land on a source of growth factor, they may elect to remain there, without interest in proceeding further to their proper target.

Why are growth factors not the "magic bullet" to cure neurological diseases? Its possible the correct dose, correct target and correct method of delivering a growth factor has yet to be discovered. "Cocktails" of several types of growth factors may be required to target available receptors. Growth factors activate key protein intracellular signals downstream from their receptors. It is conceivable that these critical growth pathways become uninterested in their presence and no longer respond.

## Playing with fire: Inflammation and damaged nerves

Damage causes inflammation. If you fall on your arm it may swell up, discolor and cause considerable pain. If you fracture your humerus at the same time, the problem is magnified, both immediately and over the next few hours! Damaged tissue inflammation develops over time and involves a series of events including our familiar blood vessel dilatation and plasma extravasation that cause redness and swelling. The areas of damage are inundated and infiltrated with blood borne inflammatory white blood cells, or leukocytes including monocytes that become tissue macrophages. Activated macrophages have wide actions that include generating inflammatory molecules such as free radicals, hydrogen ion, and proteolytic proteins. Release of immune signaling molecules calls for reinforcements to the area of injury. The inflammation that comes from "closed" tissue damage is "sterile" involving cell disruption but without invading bacteria or viruses. The immune system is activated by cell damage and the release of cellular contents normally hidden behind the cell membrane.

Nerves, especially large nerve trunks that contain many axons, undergo a form of sterile inflammation when they are injured. This process closely overlaps with the process of AxD discussed earlier. Neutrophils, the most aggressive of the blood borne inflammatory immune cells, are early invaders into sites of nerve injury.[147] Macrophages take a few days to accumulate in the injury zone of the nerve but then are expected to leave the site of damage when their work is complete. Delays in both their appearance and their exit have impacts on regeneration.[148] Equally important is the large selection of molecules that macrophages add to the process including growth factors and inflammatory cytokine and chemokine molecules.

Cytokines are small proteins that signal from cell to cell, and are largely released by inflammatory cells. However some cytokines are released by non-immune cells. It seems everyone these days wants to be an immune cell! Unique circumstances during inflammation change the behavior of cells we thought we understood. For example, a SC cell might synthesize a cytokine! Overall a flood of cytokines appear in damaged nerves many of which are pro-inflammatory, but others are anti-inflammatory. Their actions require individual and specific receptors. Interferons, first discovered in the 1950s, are cytokines that "interfere" with viral infections. Interferon family members also include one of the oldest and most established forms of therapy for multiple sclerosis (MS). Interferon β-1b, known as betaseron, is anti-inflammatory, a property that explains its benefits during the inflammation of MS attacks. In contrast, another family member called δ-interferon, when given to MS patients during clinical trials several years ago, was found to worsen the disease! These findings have highlighted that some cytokines reduce inflammation but others worsen it. Betaseron is not effective for peripheral nerve inflammation.

Other cytokines include the large family of "Interleukins" that also impact inflammation. There are at least 36 subtypes! Interleukins are best known for activating other immune cells, helping them to mature and proliferate. They also attract several types of immune cells to move toward a zone of inflammation (chemotaxis), support antibody production and more. Like interferons, some members worsen inflammation and others lessen it. IL-1 (interleukin-1) is closely associated with fever that develops after infection. IL-6 directly impacts nerve regeneration, an interesting finding discovered by Drs. Patricia Murphy and Peter Richardson, a neuroscientist and neurosurgeon respectively from Montreal.[149] Their work established that Il-6 was synthesized by neurons and that its levels increased if the neuron was injured. Moreover, IL-6 protected neuron cell bodies from dying after an axon injury.

Tumor necrosis factors (TNF-TNFα or TNFβ) have an interesting and provocative name. The TNFs are important cytokines produced by macrophages but also by other immune cells and even neurons! Classified as proinflammatory, they can overwhelm or destroy tumor cells and other cells. Despite being released by neurons, TNF can also destroy them. Three American investigators are credited with their discovery: Granger at UCI in 1968, Ruddle at Yale the same year, and Old from New York in 1975 who coined the TNF name [https://en.wikipedia.org/wiki/Tumor_necrosis_factor_alpha]. TNFs bind to their receptors, TNFRs (TNF receptors) and activate a complex "death signaling" pathway. This pathway includes destructive molecules called TRADD (Tumor necrosis factor receptor type 1-associated DEATH domain), TRAF (TNF receptor-associated factor 1), RIPK (an appropriate name for a death protein, but it stands for receptor-interacting protein kinase) and caspases. Caspases, mentioned earlier under AxD, degrade other proteins by attacking their cysteine amino acids. Just to be difficult, one protein activated by TNFs subverts the process by having an opposite, anti-inflammatory role. This contrary protein called NF-κB (nuclear factor kappa-light-chain-enhancer of activated B cells), moves into the nucleus and helps the cell survive, rather than die. Think of it as the sole model student in an unruly class who goes to a higher authority (the principal; nucleus) to report what is going on and to restore order. Behavior suddenly improves! Neurons also benefit from NF-κB's prosurvival actions.

Chemokines are subtypes of cytokine that are about half their size or smaller. Their function is to entice cells to move toward them, a process mentioned earlier and referred to as chemotaxis. CCL, CXC, $CX_3C$ are some of many family members and they in turn have several subtypes. For example, 27 members are in the CCL family. Other older names include MCP-1 (CCL2), RANTES (CCL5), MIP1a (CCL3). Please do not attempt to memorize their names! Chemokines are released by macrophages and other cells and signal other inflammatory cells to "come hither," ultimately enlarging and widening the inflammatory response. Neurons and axons begin to die if the response is overly intensive. Chemokines activate signals that change the behavior of surviving neurons.

I will briefly introduce several ways to kill a cell. Necrosis is a rapid and inescapable form of death. It is almost instantaneous not unlike being killed in an explosion, or shot through the heart. The cell goes out with a "bang." In the nervous system, a stroke, depriving areas of the brain to blood flow and oxygen causes neurons to die by necrosis. Necrosis can also follow severe insulin reactions, and some poisonings.

Apoptosis is a more thoughtful, progressive or tortuous process, depending on your point of view. Unlike necrosis, it requires protein synthesis to develop. The targeted cell generates a set of new but lethal "death" proteins within itself, a process is called "programmed" cell death. "Apoptosis" comes from the Greek referring to "falling off," like leaves gradually falling from a tree. Some of the death proteins include activated caspase-3 described above that severs cysteine amino acids of proteins, Bax which collaborates with mitochondria to kill cells, cytochrome C (a mitochondria protein), Bad (BCL2 associated agonist of cell death), other caspases and a whole range of additional characters. You will notice that inappropriate involvement of mitochondrial proteins is an important feature. Mitochondria are essential for the delivery of energy to cells. Paradoxically however, their proteins also participate in programmed cell death. Over hours or longer, apoptotic cells develop nuclear shrinkage, blebs or outpouching of the cell body then eventual fragmentation of the cell.

Why should cells have inherent "suicide" programs built within them? One important role for apoptosis is to destroy cells turning cancerous, a process that is encouraged or enhanced during cancer therapies. In the case of neurons, apoptosis may be a way of liquidating unruly but abnormal neurons that disrupt the neural network. An equivalent series of anti-death proteins are also produced in neurons and other cells that resist apoptosis. The overall balance between proapoptotic or antiapoptotic signals in a given cell determine whether it will live or die. A virus might want to enhance survival since the invaded cells produce hundreds of copies of the virus. After the virions are released, the cell becomes dispensable. Alzheimer's disease kills neurons slowly, over years, by apoptosis. In ALS motor neurons in the spinal cord die by apoptosis.

Necroptosis is a hybrid of necrosis and apoptosis that happens quickly, is promoted by TNF but requires the cell to make a few selected proteins including RIPK! "Senescence" of neurons is another form of cell demise. It describes a slow dwindling of its capabilities over time. The relative roles of apoptosis and senescence in neurons is not well understood.

Surprisingly, we know relatively little about peripheral neuron cell death. It is also important to distinguish cell death from axon death or AxD. To be clear, AxD does not imply that the neuron is destined to die by apoptosis, necrosis or necroptosis. Apoptotic retrograde death of neurons does occur after axonal damage, a topic we have discussed earlier, but only in a subset of neurons. Loss of the axon might be comparable to having a leg amputated, an outcome that is difficult to accept but not fatal. In the case of cell body death, the entire neuron "tree" degenerates and disappears, including both perikarya and axon. It sometimes may become difficult to distinguish early apoptosis of the neuron from AxD. For example, as the neuron slowly degenerates from apoptosis, its distant axon terminals may be the first to show signs of degeneration, resembling AxD. The process may initially appear to be an axonal disorder but more serious consequences follow. The distinction is important because axons can be regrown but replacing entire neurons is challenging.

Cytokines like TNF can damage neurons or sometimes protect them, depending on their dose and the local environment. If a barrage of cytokines appears on the horizon, the neuron or axon is not destined to do well. If the neuron is exposed to only a few cytokine molecules a few benefits such as enhanced regeneration are possible. For example intense inflammation of sites of nerve repair, such as an infected tube or conduit connecting transected nerve stumps, completely block growth.[150] However, adding low doses of selective cytokines help nerve repair. While nerve trunks have their own "resident" macrophages, many more enter from the blood stream after injury, along a well recognized timetable. For example, influx typically occurs at approximately 5–7 days following injury. If macrophages are denied entry regeneration proceeds more slowly. Finally, we have learned that macrophages come with different personalities. Like cytokines, some are proinflammatory, called M1 and are generally unhappy characters whereas others, called M2 are anti-inflammatory and support growth and repair. Several in between varieties also exist. Like people, macrophages can have good M2 days, and bad M1 days. Once the nerve is regenerated however, they are no longer welcome and may become a problem, an interesting development described by Dr. Sam David at the University of McGill.[148] One downside of the inflammation experienced by nerves is its impact on the development of pain after injury or in neuropathies.

## Neuron HQ central and damaged axons

What would happen if your leg was amputated? It is unlikely you would be happy about it. You might experience severe pain and complain loudly if an anaesthetic had not been given. Its very likely that you would never be the same person unless the leg grew back fully! Here we consider neurons that have had injury to their axons. While the injury might be some distance away from the neuron cell body, it will not be happy about the state of affairs. Does the neuron react or complain? We have discussed neuropathic pain and the "ectopic" or abnormal discharges set off by a neuron that is injured. Unlike your leg, a severed axon can be replaced, restoring the neuron to some semblance of normal. Before this happens however, the perikarya changes in a dramatic fashion, despite how far away the injury occurs. Something has happened "out there."

A wave of calcium is the signal that an injury has occurred. Once the integrity of the axon is breached by a crush or cut injury, this wave spreads along the axon and reaches the cell body. At the perikarya, the calcium signal changes what genes are active and how neuron proteins function. For example within an injured axon, the proteins vimentin and importin transport a signaling protein called Erk directly into the nucleus of the neuron.[151] Erk then turns on key genetic signals called transcription factors found on the nucleus that interact with the DNA of the neuron. Transcription factors activate or suppress specific portions of DNA responsible for transcribing mRNA and thereby the proteins it produces. Erk is thought to be a general transcription "switch" turning on a number of key DNA programs including those responsible for growth. In this way, the neuron learns to behave as a regenerating cell, moving beyond its function as simply part of the wiring.

Some neurons cannot withstand the axon amputation and die, especially if they are embryonic or neonatal neurons. One of the first examples of neuron apoptosis was discovered in embryonic neurons that underwent an axotomy, simply another term for an axon injury such as transection. Massive retrograde loss of embryonic neurons was prevented if their cut ends were supplied with a growth factor like NGF. This remarkable finding established that access to growth factors was essential in keeping neurons alive. Axotomy interrupts the capture and retrograde transport of essential growth factors to the cell body. This idea is the basis for the "neurotrophic hypothesis," the understanding that neurons depend on a supply of growth factors to keep them functioning. When a tissue is innervated by several axon branches, limitations in the availability of growth factors determine how many branches will survive. While there is strong support for the neurotrophic hypothesis during developmental stages, this is not the case in adults. Fortunately adult neurons are more robust, have found a way to compensate for such catastrophic loss and experience less retrograde death, if any. It is a controversial point however and later loss, albeit limited, may complicate adult axotomy. In contrast, 90% retrograde loss would be catastrophic eliminating hope for significant regeneration. There would be no reason for me to write this section if such loss were expected; the topic would instead be about neuron replacement. While we will discuss "new neurons" later, in most instances perikarya survive, ready to fight another day by sending out new axons.

Neuroanatomists discovered that neurons change in appearance after they have had an axotomy. Termed the "cell body reaction," the most common change is a shift in the location of the nucleus from the center of the neuron to the side.[152] Why is this so? Axotomy alters the content of lattice proteins in the neurons including their content of neurofilament, that maintain the shape of the neuron and hold the nucleus in place in the center of the neuron [personal communication V. Verge]. As neurofilament levels decline after axotomy, the position of the nucleus drifts to the side of the neuron. The color properties of the neuron may also change. Traditional stains used for neurons called hematoxylin and eosin (H + E), normally identify a blue tinge in their cytoplasm arising from their content of ribosomes (cellular structures that help to synthesize proteins from RNA) with their associated RNA. After axotomy, the stains identify a remarkable shift of this staining from the cytoplasm throughout the cell to the inner edges of the cell membrane. This striking alteration is among that most easily recognized in the "cell body reaction."

After axotomy, surviving adult neurons dramatically change their outlook on life. Their behavior in an intact nervous system is active but stationary. These neurons and their axons stay healthy, transmit impulses, obey the constraints of their life and fufill their role in it. Their actions are characterized as a "stable" transmitting role and part of the network. However post axotomy injury, a series of new proteins are synthesized and the neuron switches gears to a "regenerative mode." Valerie Verge at the University of Saskatchewan was among the first to detail these changes in sensory neurons, helping us to understand "regeneration associated genes" or "RAGs."[153] RAGs are genes coding a series of proteins that undergo heightened synthesis after an axotomy injury. Alternatively, some RAGs are tuned down by axotomy. Classical upregulated RAG examples include beta-3-tubulin, an essential protein for microtubules required for active growth cones and for fast axoplasmic transport. Growth associated protein-43 (GAP43) is another RAG that was simultaneously discovered by an American scientist Pate Skene and Dutch Canadian neurochemist Henrik Zwiers (who called it B50).[154,155] Three other laboratories also participated in its discovery. In the end the American side won the name (GAP43 stuck, B50 did not). GAP 43 is exported from the nucleus down the axon by anterograde axoplasmic transport to support the growth cone. Unlike the upregulated RAGs, the genes that make neurofilament proteins reduce their synthesis after axotomy and allow nuclei to shift their position during the cell body reaction. The downregulation makes sense because neurofilaments are not essential for the initial stages of growth. They are required later, as the axon matures and acquires girth.

The repertoire of known RAGs has dramatically expanded with molecules named ATF-3 (cyclic AMP-dependent transcription factor), c-jun and many others. New methods such as deep RNA sequencing have identified more RAGs than scientists available who could study them. Many require a full scaled investigation to clarify their impact on neurons. One word of caution is in order here. One might assume that to be important in regeneration, a protein associated with a RAG should have either increased or reduced expression after injury. This presumption may be incorrect. Several proteins in sufficient amounts may already be available to influence regenerative success. Thus, the change in activity of a RAG and the levels of its protein during regeneration do not necessarily indicate its importance. Instead these changes represent clues that a RAG might be relevant for growth. When neurons reach their targets and resume a normal existence, RAG levels normalize their synthesis of proteins. In fact, some RAGS resume normal activity sooner. For example, adult axons that have not yet reached their target may enjoy a close relationship with their partnering SCs. These "reactive" SCs, growing and migrating after injury, move with axons, support them with growth factors and are good friends or partners. In doing so they may signal to axons and neurons that RAG expression is no longer required. Perhaps this explains why adult neurons largely survive after axotomy unlike embryonic or newborn neurons. SCs are a source of support after axotomy that other neurons may not be benefit from.

Is it possible to push RAGs further to improve regeneration? Investigators have considered this interesting idea. Dr. Tessa Gordon and colleagues discovered that electrically stimulating an injured nerve persuades it to think more deeply about regenerating. In exhaustive work that examined multiple approaches to stimulating injured nerves, this group made a remarkable discovery. After a nerve is transected, a short period of stimulation to the proximal stump of the nerve at 20 Hz (20 stimuli every second) given for only an hour has a profound impact on regeneration. Growth of both motor and sensory axons is dramatically improved.[120,156] The protocol is now termed ES, for electrical stimulation. ES operates by pushing RAGs to earlier and higher levels. Several laboratories, including our own have confirmed the impact of ES in various animal models. In fact, if you remove adult rat DRG sensory neurons and harvest them, "in vitro," some of the growing neurons can be placed directly on top of and in contact with stimulating electrodes. Neurite (axon) growth from neurons directly stimulated is dramatically different from those nearby but not stimulated.[157] With Gordon, Ming Chan and colleagues at the University of Alberta have translated this work to persons undergoing nerve repair, completing some of the first clinical trials worldwide of a specific intervention to treat nerve injury.[158,159] So far, the research has identified benefits of ES in three separate types of nerve damage: carpal tunnel syndrome, ulnar neuropathy and injuries of digital nerves in the fingers. The work in humans is ongoing and exciting.

"Translation" is the current mantra of agencies that fund research. It literally refers to whether discoveries in the laboratory can be "translated" into a real and meaningful result in the health care domain. Politicians might also insist on knowing whether the treatment saves money. The expectation is that experimental work finds a role in improving human health. However, some research funders have expected this sooner than later! Does this work? If a funder invests $1M into research but demands that it results in a new drug, perhaps for a specific form of neuropathy, will there be an early return? The interesting fact is that "targeted" research fails much more often than it succeeds. Most scientific discovery has been unfettered and "game changing" discoveries in biology and in science generally have not followed a plan.[160] Discovery is not by nature predictable and if it were, results might be suspect! Despite these caveats, inspiring examples of basic science discovery leading to immediate patient benefit have occurred. In our discussion of ES in regeneration, biology and experimental work had a rapid win. Collaboration and team work were responsible: Tessa Gordon and Valerie Verge, neuroscientists, Ming Chan a physical rehabilitation specialist, Tom Brushart and Jaret Olsen, nerve surgeons. The role of ES after nerve injury requires further investigation. For example, how long after an injury can ES be applied and will it be effective then? Are several treatments ES more effective than a single application? Finally, does ES have an impact beyond enhancing RAG activation? BDNF, the second member of the neurotrophin growth factor family may be a key signal that is activated by ES and that triggers RAGs.

Whether there are other ways to ramp up RAGs to improve regeneration is not known. My expectation is that this possibility is achievable. Like ES, many questions persist such as how long RAGs should remain activated? If resurrected a little later after an injury, will regeneration get another jump start? Is it better for nerves if the RAGs get turned off at some point? What are the important lines of communication between RAGs in cell bodies and regrowing axon tips?

## Are there new neurons?

Possibly. One of the most important discoveries of $20^{th}$ century neurobiology has been the identification of stem cells in the adult central nervous system. Found in small and specific repositories of the brain, stem cells can differentiate into all of the cell types in the brain, including neurons. They are found in the dentate gyrus, subventricular zones, areas beneath the ventricles (fluid filled chambers in the brain) and the olfactory bulb (responsible for smell). Discovered by Drs. Sam Weiss and Brent Reynolds in early 1990s,[161] detecting their existence in the CNS represented yet another Canadian contribution to neuroscience. There is an ongoing debate about whether human stem cells continue to generate significant numbers of hippocampal neurons involved in memory during adulthood.[162]

The topic of stem cells has received only limited attention in the PNS. Why is this the case? I have already made the point that PNS work, despite its importance, remains "below" the radar, overshadowed by studies of the CNS. Many investigators believe that the PNS might not require a source of stem cells at all and they may be correct. Many disorders of the PNS do not involve loss of parent neurons. Their numbers, in ganglia or anterior horns, may be perfectly adequate and the real problem and associated disability may arise from damaged axon wiring.

In some instances however, peripheral neurons are destroyed. This loss involves not just the axons or their terminals but the entire neuron tree including the perikarya, or cell body. With such a catastrophic state of affairs, an infusion or transplantation of new stem cells to replace neurons might be a useful strategy. Successful replacement of neurons however, would pose further challenges. New neurons would be tasked with sending new axons down the entire length of the nerve to an appropriate target. Catastrophic PNS disorders with neuron loss are called "neuronopathies" as opposed to "neuropathies" because the target is thought to be the cell body but not the axon. In reality, the distinction is artificial. It is difficult to target just one part of the neuron, and not the entire "neuron tree." For example, many neuropathies that have "dying back" of their axons may be experiencing difficulties that originate in the cell body. The cell body might not possess the resilience to withstand whatever is ailing it and as a first step may withdraw resources from distant axons. The terminals of these axons consequently retract or withdraw. Neuronopathies may be characterized by ganglia inflammation due to autoimmune disorders associated with cancer or rheumatological disorders like scleroderma. Moreover, they are thought to have differing clinical features compared to neuropathies, but sometimes the distinctions between the two are difficult to tease out. Recall that in Chapter 6 we described the deafferented woman described by Oliver Sacks and our own patient with anti-GW182 antibodies who also had severe loss of sensory neurons.[94,95,163] These are examples of advanced sensory neuronopathies.

Motor neuronopathies are well known. ALS, amyotrophic lateral sclerosis, is a motor neuronopathy in which neurons in the anterior or ventral horn of the spinal cord disappear. ALS also involves the CNS because brain motor neurons that travel to the spinal cord also degenerate. Other motor neuronopathies are a result of poliomyelitis, West Nile virus or other enterovirus infections. Spinal muscular atrophy (SMA), untreatable until recently, is a motor neuronopathy that has a genetic cause and can involve infants, children or adults. Research work has discovered mutations in a protein called SMN1, short for "survival motor neuron" that cause the demise of motor neurons in SMA. The complex genetics of SMA involve partial compensation for a dysfunctional SMN1 protein by a sister protein SMN2 that is less adept at supporting motor neurons. The severity of the disease depends on how well the sister protein can compensate. There is simply not enough time to pursue all of the remarkable questions in PNS neurobiology that have emerged! Some have already led to dramatic new therapies such as investigations into the relationship between SMN1 and SMA. Its treatment has not involved neuron replacement. It is, however, clear that PNS disorders can have either neuron loss, axon loss or loss of both. Given this, replacement by new neurons may be welcome in selected disorders and circumstances that involve parent neuron loss.

New neurons may be able to arise from stem cells located in the center of the spinal cord, specifically from its "central canal." This small fluid filled chamber runs the full length of the cord, is abnormal in syringomyelia discussed earlier but may harbor a small population of stem, or neuron progenitor cells. At this time it seems unlikely that they can transform into motor neurons in sufficient numbers then migrate to the anterior horns where they are required. When motor neurons disappear during ALS, aggressive new neuron formation would be essential to compensate. At the time of writing, some physicians have started clinical trials that involve infusion of stem cells to replace lost motor neurons in ALS patients. Time will judge the success of this strategy. The promise of "stem cells," frequently from clinics with questionable expertise, has led persons with serious neurological disorders to travel the globe seeking a treatment. Most of these "stem cell" clinics however have little evidence to support the proposed benefits. Buyer beware.

Dorsal root ganglia may also be capable of generating new neurons. While recently verified, the idea emerged earlier, and prior to the discovery of adult neurogenesis. A group of investigators from Israel reported that numbers of carefully counted DRG neurons increased over time.[164] However, these reports were met with a wave of skepticism since total numbers of DRG neurons are surprisingly difficult to count. The DRG is a three dimensional structure not unlike an egg and how the nerve is cut/sectioned and what you count when you examine neurons can be complex. Needless to say, the critics of this work insisted that the counts of DRG neurons were flawed and that DRGs were not capable of neurogenesis.

When stem cells were discovered in adult brains by Weiss and Reynolds in 1992, they formed interesting and characteristic balls of cells called "neurospheres" after removal from their niche.[161] Neurosphere stem cells are more properly called "multipotent" because of their capability to form neurons and glial cells, the cells of the brain. Another term used for them is "precursor" cells. "Pluripotent" stem cells found in embryonic tissue are somewhat different and possess the ability to form any kind of cell, beyond brain cells alone. "Totipotent" cells can not only form any cell in the body but they can also form placental tissue, indicating that they are yet more fundamental. When multipotent, pluripotent or totipotent cells develop into specific cells types like neurons or glial cells, this is called "differentiation." In a sense, they are converted into adult cells. Support for the idea that DRGs can generate new neurons followed the discovery that they can generate neurospheres.

What is the purpose of stem cells in DRGs? Do they contribute or simply sit there to fascinate biologists? Perhaps they can be encouraged into action. For example, it may be that a series of molecular signals can persuade DRG stem cells to differentiate into new adult cells such as neurons, glial Schwann cells or others. This might be beneficial or harmful depending on the cell type. Stem cells might form satellite cells found around neurons or SCs supporting outgrowing axons from the DRG. Along these lines, it is known that the glial cells surrounding neurons, the perineuronal cells, are in a constant state of flux. This interesting form of replenishment, involving stem-like cells that express the protein Sox2, was identified in recent work by Anand Krishnan in our laboratory mentioned earlier.[144] Stem-like cells are found around neurons in close proximity where they may be about to form glial cells and perhaps others. When an axotomy occurs, the perineuronal glial cells proliferate and the stem-like cells persist. Perineuronal glial cells also die more readily than neurons. Thus, in this DRG hubbub of birth, reproduction and death, is there an ongoing role for stem cells and do they ever form neurons?

How important might it be to generate new SCs from stem cells? Interestingly, SC precursor cells can be grown from the skin. These are called SKPs, or skin-derived precursor cells discovered by Freda Miller, Jean Toma and colleagues in Toronto.[165] They behave like activated SCs that can migrate, proliferate and synthesize large amounts growth factors and other molecules.[166] Given the critical roles of SCs following nerve injury, it is possible that adding SKPs will help, especially if there is a large gap between the proximal and distal stump.[167,168] This possibility is an exciting new approach to nerve repair. Recall however that SCs proliferate on their own after injury in any event, so whether SKPs are also needed is unclear. Too many SCs added to the mix might also be a problem because, like neurons, they may compete for resources. Uncontrolled proliferation of SCs can also cause benign, and rarely malignant tumors called Schwannomas.

There is an instance where transplanting SCs may have obvious importance. In patients with severe inherited CMT neuropathies, discussed above, the SCs that carry the genetic mutation lead to the disease. If we eventually identify a way to treat severe CMT, one option might include infusing new robust SCs that contain all the genes they need. Using new techniques of gene editing, it may be possible to use the patient's own SCs or stem-like cells to generate these fresh intact SCs for replacement. Another approach might be to simply correct the genetic defect of their existing SCs in situ with gene therapy. Replacing SCs by injection might be an impossible task considering the necessity for widespread distribution.

## What is collateral sprouting

We have made the point that axons are not content to stagnate. They have ongoing growth or plasticity unless they reach a target that insists that they pause. Once they reach a neuromuscular junction, it is unlikely that they will keep sprouting or moving. Better to stay put! In the skin however, we have argued that they are in flux because their neighbors, skin cells or keratinocytes, are always on the move.

Axons seem to sense the presence of their neighbor axons. Firstly, they avoid invading the same territory of skin. This may involve repulsive proteins that tell their neighbor to stay away from their turf. Flies have a group of proteins called DSCAM (Down syndrome cell adhesion molecule), that exhibit tremendous diversity (> 18000 distinct types)[169] through minor changes in their structure. In doing so, they provide "barcodes" for axons allowing self to recognize self or nonself. Through self recognition axons may be able to repulse neighbors that wander too close. In mammals including humans other molecules may instead have this role. Our body spaces its nerves in a fascinating manner- not too many, not too few. More nerves must be positioned at the finger tips, less over the back. Its an orchestra and every player knows where it needs to sit and why. However, we don't really know why!

Secondly, axons seem to sense that a neighbor has died. Their interest might be motivated by empathy or avarice. As a charitable response, axons send out extra branches to compensate for what was lost when their neighbor disappeared. In this way, axons that have managed to survive can prevent loss of sensation. The rescued sensation might not be as rigorous or detailed, but nonetheless there are branches where none might have survived before. The response whereby a surviving neuron sends out a branch to cover off the territory of another is called "collateral sprouting." This process can be vigorous and it differs from sprouting of damaged axons during regeneration. Overall, collateral regeneration may offer surprisingly large compensations for the loss of neighboring axons. Why is this considered avaricious or greedy? Recall that in developing nervous systems, axons and neurons compete for growth factors. Those that are not successful simply die, a Darwinian outcome. This is less drastic in adults but it may be that extra branches of neurons invade vacated property to collect unused growth factors. No longer repulsed by proteins on their neighbor axons, they pillage at will. It's a situation that is helpful for replacing loss of axons but also exposes uncharitable biology.

We have learned that a regenerative sprout arises from a damaged axon but a collateral sprout comes from an intact axon. In fact, collateral sprouts are thought to arise from Nodes of Ranvier. While it is not clear why this occurs, Nodes of Ranvier might offer a clear passage that avoids myelin as they travel outward. Myelin sheaths might inhibit sprout formation. What would persuade perfectly fine and apparently happy skin axons to suddenly send out branches into unknown territory is worth thinking about.

In the skin, collateral sprouting can cover a sizable amount of territory. One of the most striking examples of this is in humans who have had a sural nerve removed. This small nerve behind the ankle arises from behind the knee, runs down the back then outside of the lower leg, around the ankle bone (lateral malleolus) then into the side of the foot. The sural nerve and its branches are responsible for sensation on the side of the foot up to and just above the ankle. It is sometimes removed for diagnostic biopsies leaving an area of dense numbness. Since over 7 cm of the nerve may need to be removed, it is unlikely the nerve will regrow to the original site, and a neuroma forms at the proximal stump. Theriault and colleagues[41] mapped the territory of numbness over the skin after a sural nerve biopsy by drawing a map of the numb portions and calculating its area. They waited 6 months to a year later and remapped noting that the area of numbness had dramatically diminished. However, during recovery sensation returned from the edges in, and the sural nerve itself had not regrown from its proximal stump. Collateral sprouting accounted for the recovery. Sprouts from nearby intact nerves, such as the superficial peroneal nerve and the saphenous nerve, neither damaged by the biopsy, came to the rescue, or the plunder depending on how you look at it.

Jack Diamond, another Canadian neuroscientist, described interesting biology concerning collateral sprouting. He observed that collateral skin sprouts required NGF.[170] For example, providing antibodies against NGF prevented sprouts from forming. In contrast, such antibodies did not impair regenerative sprouts from forming after a nerve injury. While Diamond's findings have been debated, they have not been disproven. Little new work has emerged over this interesting problem.

Do motor nerves have collateral sprouts? Yes. We have mentioned this development earlier. Clinical neurophysiologists have long recognized that collateral motor branches may be more common and important than sensory ones. On an electromyogram (EMG), activated motor units that are recorded from muscles may be enlarged, sometimes dramatically, as a hallmark of motor collateral sprouting. EMG electrodes within muscles listen to and display the electrical activity of motor units when a muscle is voluntarily contracted. Since the motor unit is a single motor axon connected to a number of muscle fibers, its size tells you how many muscle fibers it connects to. For example, persons having had prior poliomyelitis may lose a large proportion of their motor neurons and axons. However, over time this loss is compensated for and muscles from persons with prior poliomyelitis develop giant motor units. Remaining motor axons sprout to denervated neighbor muscle fibers and reduce the impact of the loss of nearby motor units (Fig. 19). When muscle fibers are deprived of their motor nerve connection they develop wasting, also known as atrophy. A collateral sprout from a good (or greedy) but healthy neighbor is all it takes to keep the abandoned muscle fiber from wasting. Motor unit enlargement can therefore be diagnostic by identifying whether weakness arises from loss of motor axons or other causes such as CNS disease or primary muscle disorders with very different EMG activation. The final part of this story is that vigorous motor axon collateral sprouting can obscure how extensive the underlying loss of motor neurons or axons is. This can be the case in ALS in which motor neurons slowly disappear over months. It is estimated that we can lose substantial proportions of our motor neurons before we notice weakness! This may also be the reason that loss of muscle strength with aging can appear to be slow despite steady age-related loss of motor neurons. Of course if you do not exercise and train, you may lose strength for completely different reasons, namely deconditioning!

## Persuading neurons not to be boring

Most neuroscientists would cringe at the idea that neurons might be boring. They are not. However, classical teaching is that they merely stay in place. Neurons and axons are designed to be part of an electrical grid. Just as a stray piece of wired just might short circuit a whole network, unwanted new connections from axon sprouts should be suppressed vigorously! So, in some sense of the word, neurons are a little boring. The challenge comes when they are asked to do something unexpected and extraordinary: to grow after an injury or to collaterally sprout.

The beauty of research is that findings from a completely separate field can provide exciting answers to the area you are studying. While there is power from focus, the danger is that you may not realize the answer is a department or research group away! In the case of neurons, there have been no clear explanations of why neurons might be boring, content to be a piece of wiring in a puzzle. In other words, why do they stop growing? Older ideas have fixated around interactions. In this explanation, their targets fully provide neurons with support and there is no need to be promiscuous. This might involve adhesion and basement membrane molecules, growth factors and other ways to keep connections intact. In fact, a classical tenet of neuroscience is the Hebbian hypothesis alluded to earlier in the book, named after a Canadian psychologist, Donald Hebb (1904–1985). His hypothesis states that the "neurons that fire together wire together." In this light, two neurons that connect and fire together strengthen their connections and do not sprout elsewhere. The Hebbian hypothesis has considerable merit and is supported by extensive experimental work.

Neurons may be programmed to be boring. In a separate field, cancer biology, a group of molecules called "tumor suppressors," when inactivated by a mutation, give rise to cancer. Uncontrolled cell growth develops because a "check" or brake on growth has been eliminated. Normally the nonmutated version of the protein, the "tumor suppressor" inhibits key growth pathways by blocking growth molecule synthesis, or by other means.

Our genes made from DNA are decoded by the process of transcription into mRNA, which then undergoes translation into proteins. Some proteins are called transcription factors because they enable this DNA decoding into RNA. Transcription factors are proteins that participate in mRNA production and help it to succeed, or alternatively shut it down. Some transcription factors are specific to the production of one certain kind of mRNA and protein. Others have wider actions, such as influencing a large number of genes essential for life. The point of this is that some transcription factors, by signaling the synthesis of multiple key proteins can have a substantial and outsized impact on overall growth. Thus you might easily imagine that for all the proteins required for a cancer to grow, a widely acting transcription factor might be critical. These could be activators, repressors or both depending on the particular genes. Tumor suppressor molecules include transcription factors among them with names like Myc, Rb1, p53 and others.

The discovery that tumor suppressor molecules were present in neurons was an important nuance. It seems that nature, once it has an exquisitely designed protein, uses it for several kinds of cells. Tumor suppressors in neurons may be critical in reminding them of the message to "curb your enthusiasm." They may exist to help preserve the fidelity and exquisite biology of neural connections. Suppressing unwanted growth is important to preserve the wiring.

A decision over whether a neuron is stable, boring or in a growth spurt may depend on the amount and distribution of growth suppressive protein within it. The normal nervous system may not take into account the possibility of a serious injury or need for new growth. While we have discussed how nerve injury upregulates RAGs, how does it handle growth suppressive proteins? Some are likely downregulated, allowing growth to proceed. However others appear to be barriers to growth without a change in their overall levels. Among the first tumor suppressors discovered in adult neurons was PTEN, which stands for "phosphatase and tensin homolog deleted on chromosome ten."[171,172] The key word here is phosphatase. Many proteins, including those of the nervous system are activated by attaching a phosphate molecule to them by enzymes called kinases. It is an important, common and widespread action that spans almost all forms of life. Of course, when you add something with an enzyme, you typically need another enzyme to remove it. The enzymes that remove phosphate molecules from proteins are called phosphatases. PTEN is such a phosphatase and its actions alter the behavior of proteins in a critical growth pathway. As such, kinase and phosphatase molecules are central to how almost all growth factors signal neurons.

Just beyond the molecular step in which PTEN curbs growth pathways by being a phosphatase, is a molecule known as pAkt. pAkt is essential to many types of growth and for survival of neurons. While not a direct transcription factor, pAkt signals widely in the cell to support growth pathways. The p in pAkt denotes the attached phosphate molecule and indicates that its interaction with a phosphatase, capable of removing it, might not benefit neurons. Although there are a couple of intermediary steps in the pathway not described here, PTEN activation inhibits Akt from becoming pAkt and shuts down growth. This is an example of how a single protein can have a wide influence. Its impact does not depend on whether signals originated from NGF, BDNF, insulin or others. PTEN operates within the neuron "downstream" of growth factors as an internal or "intrinsic" feature the neuron. Stopping or removing PTEN makes neurons far less boring.

That PTEN suppression increased growth was first discovered in the CNS by Park, He and colleagues using retinal ganglion cells in the eye tasked with sending axons into the optic nerve.[171] In work from our laboratory, Christie et al demonstrated the remarkable impact of PTEN inhibition or knockdown in peripheral neurons.[172] She observed that adult DRG neurons, plated in a dish, had enhanced growth if PTEN was inhibited with a compound that prevented its phosphatase action. Another strategy for inhibiting the effects of a protein involves suppressing its translation from mRNA. New molecules known as siRNAs are small versions of RNA that bind to and block the messenger RNA (mRNA) that is used to form a protein. "Knocked down" is a term biologists use when they inhibit a protein from being synthesized by using an siRNA, or related molecules known as shRNAs, morpholino RNAs or ASOs (anti-sense oligonucleotides). All of these operate in a similar way. Using siRNAs, Christie showed that knocking down and inhibiting PTEN achieved the same effect. Not only were "normal" adult DRG neurons activated to grow better but, neurons that were "preconditioned," a state of enhanced readiness to grow, described earlier, grew yet more vigorously. For the first time, we demonstrated that a preconditioned neuron, already ramped up to grow aggressively, could be coaxed into further growth. Finally, PTEN inhibition and knockdown also increased the number of axons growing out from an experimental nerve transection. The strategy also worked "in vivo," on a nerve from a live rat.

Unlike neurons in the body, "in vitro" neurons send out branches in all directions. Using the technique of immunohistochemistry that labels neuron proteins such as neurofilament or βIII tubulin, you can visualize the cell body and all of the processes extending out and creating branches. The processes are spectacular. To make the point, a former graduate student in our laboratory, Shane Eaton, displays and markets images like this to interested buyers! If you look at enough neurons in vitro you will realize that some are round cells without branches and others have a literal forest of branches extending from them. Why some decide to be exuberant and others less no is not clear. We have already discussed how the sensory ganglia contain several kinds of sensory neurons. We now believe that some are built to grow more slowly and others quickly. One of the fascinating aspects of PTEN is that its content differs among cell types. Higher content of the phosphatase was noted in smaller neurons called "IB4" or nonpeptidergic (also called Mrgprs earlier), different than the SP peptide containing neurons. Lectin binding IB4 neurons grow very slowly in vitro, a discovery by Tucker and Mearow at Memorial University in Newfoundland.[173] In other words, one might directly connect the growth potential of some neurons with their PTEN content. This does oversimplify the matter, since PTEN levels might be one among several master growth switches. Several others exist.

A final twist on the PTEN story followed when a former graduate student in our laboratory, Bhagat Singh discovered that sensory neurons from diabetic mice had higher levels of PTEN.[174] Diabetes is discussed separately but recall that it is associated with slowing of nerve regeneration. When Singh knocked down PTEN by siRNA in these studies he could restore regeneration in diabetic nerves to normal levels.

PTEN inhibition and knockdown have gained the attention of neuroscientists interested in regeneration both in the CNS and PNS. In the PNS, we still do not know whether inhibitors of PTEN could be given for prolonged periods to improve regeneration outcomes. Long term inhibition of a tumor suppressor molecule might be a problem, inadvertently promoting tumor growth. It may be that very limited forms of short term PTEN inhibition or knockdown strictly at a site of nerve injury might avoid the risks of cancer forming in nearby tissues.

Not satisfied that PTEN represents the only growth inhibitory protein in nerves, our UofA laboratory has pursued several others. Each has unique properties that allow a them to behave like PTEN. Recall that some transcription factors activate very widely in cells. Rb1 stands for retinoblastoma protein 1. When mutated it can cause eye tumors in children known as retinoblastomas. These are highly malignant tumors that require removal of an eye or other therapies to prevent their spread. Rb1 inhibits a key transcription factor known as E2F1 that acts extensively on DNA with actions that support growth. When Rb1 is mutated, as in the childhood eye tumor, it no long restrains E2F1 and enhanced growth occurs.

As with PTEN, nonmutated Rb1 was identified in most peripheral neurons of ganglia.[175] There has been no known function for Rb1 in neurons except perhaps to "curb enthusiasm" as above. Knockdown of Rb1 using siRNA also dramatically increased the arbor of neurites around sensory neurons. The images were as beautiful as those with PTEN inhibition. Moreover, the acceleration of growth occurred whether the neurons were "primed" by preconditioning or not. Rb1 knockdown also improved regeneration of nerves in rats or mice with a cut or crush injury of their nerves. No human studies have been undertaken.

A host of other important proteins restrain or influence growth as established in tumors. Our laboratory has new evidence that APC, which stands for adenomatous polyposis coli, is also a suppressor of growth in neurons.[176] This tumor suppressor was initially studied in colon cancers, but like the others discussed above, it is also present in peripheral neurons. Unlike PTEN, APC is not a phosphatase and unlike Rb1, it is not an E2F1 transcription inhibitor. APC binds to another popular and widely acting transcription factor called β-catenin and marks it for destruction by the cell. Inhibiting or knocking down APC allows β-catenin to persist, migrate to the cell nucleus and then acting as a transcription factor, widely activate DNA to promote growth. However, like PTEN but unlike Rb1, it seems to have higher concentrations in the slower growing IB4 neurons. The reasons for this more selective presence of APC are not known.

## Behaving like a baby: Developmental molecules for regeneration

A general property of the nervous system is that nerves in babies and children grow back faster than in adults. Adults are slower to learn languages, and slower to regain walking after an injury. The nervous system is an unapologetic agist, favoring the young and reticent with the older.

This prompts the idea that acquired knowledge about molecules or processes from the PNS in youth may potentially aid regeneration in adults.

Some ideas have not been helpful. For example, NGF in embryos and neonates has much more widespread actions than in adults. As the organism ages, neurons shed their NGF receptors and become indifferent. So, while molecules or pathways could simply be reused to "recapture" youth, the receptors are no longer present in the case of NGF.

On a more positive note, I will use three examples where we have learned something about adults from the PNS of youth. Maureen Condic and colleagues from the University of Utah examined the interplay between the basement membrane, the "soil" on which neurons are meant to grow and their behavior.[177] The interaction between specific proteins on the neuron growth cone and its substrate can dramatically influence growth. Laminins and fibronectin are two major forms of basement membrane protein that growth cones "see" during their movement. While both support growth, each belongs to a family of that includes several subtypes of proteins. Recall that the proteins on the growth cone that interact with the basement membrane are called integrins. Integrins have subunits named α1, β1 and others and each subunit can be of a different type. Since each integrin protein subtype composite then is slightly different, despite being from the same family, they interact differently with the laminins and fibronectins. Condic discovered that the exact subtype of integrin on the neuron has a major influence on how it grows along its substrate. Moreover, the amount of integrin on neurons changes with the onset of adulthood. When adult neurons were artificially made to produce and express (display on their surface) more "young" integrin called α1β1, these neurons grew like young dynamos on laminin, keeping up with neonatal neurons. If adult neurons were generated to produce higher amounts of another integrin called α5β1, their growth also accelerated, but this time on fibronectin. The moral of the story is that older neurons slow their growth, not unlike human beings. To make them grow better, specific integrin subtypes make a difference. This is a strategy to recapture youth in order to grow nerves better.

Shh (sonic hedgehog protein) is the name of an interesting molecule that originated with the video game "Sonic the Hedgehog." The protagonist possessed superpowers and a postdoctoral student, Robert Riddle, working in the Harvard laboratory of co-discoverer Clifford Tabin in 1993, selected this singular name [https://en.wikipedia.org/wiki/Sonic_hedgehog]. Riddle discovered the name in his daughter's reading material. Shh has several protein family relatives including Indian Hedgehog and Desert Hedgehog. It is a critical development signal that helps to structure the spinal cord. Specifically, by acting as a cell secreted signal gradient, Shh levels in the cord determine whether developing motor neurons choose the front or anterior part of the spinal cord to live thereafter. A gradient molecular signal is a continuously changing concentration that is important for development pathfinding of axons. Recall our party room analogy with increasingly friendly people. A higher gradient of a specific molecule might be attractive, drawing axons toward higher concentrations, or repulsive, pushing them away to areas with lower concentrations. Shh is also called a "morphogen" because it is responsible for the "morphology" or shape of the spinal cord. Shh has many other roles including developmental patterning of limbs and fingers. It acts, in different settings, as both an attractive and repulsive cue for growing axons.

Shh is a striking example of how a developmental molecule reappears in adults and is repurposed. Here we find Shh working its magic or powers within adult sensory neurons.[178] If you "knockdown" levels of Shh using our familiar siRNA treatment, the growth of neurons is stunted in vitro with fewer neurites. In injured nerves, Shh appears in regrowing axons but if its levels are lowered, their advancement is inhibited. Both findings indicate an important role for Shh in the regenerative growth of neurons. It appears that a series of molecules are waiting in the wings until a crisis such as a nerve injury develops. With their capabilities in demand, they emerge and help resolve the crisis. In the case of Shh, it shores up growth although its chosen collaborations are under debate.

A final example that illustrates how repurposing proteins from the young can help adults involves a group of molecules called netrin, DCC and Unc5H. All three are well known in development and hundreds of papers have been written about them. Netrin is the "ligand" or activator and DCC and Unc5H are its receptors. There are colorful full names for these: DCC for Delected in Colorectal carcinoma! Unc for Uncoordinated! Many subtypes of the receptors have been discovered, but Unc5 and in particular Unc5H are of interest to our story. To begin, netrin when released extracellularly is thought to be a cue for attracting axons. When the DCC receptor encounters netrin, it is attracted toward its source. During CNS development, these interactions are a mechanism for one neuron to connect with another. In the adult peripheral nervous system however, there are some remarkable changes. These were discovered by Dr. Christine Webber in 2011, then a postdoctoral student in our laboratory.[179] Netrin is thought to be produced by the nerve growth cone but its receptor DCC is instead found in SCs and after injury, its levels rise dramatically. Leading SCs emerging from an injured nerve in search of a distal stump express DCC and are receptive to ambient netrin. This netrin-DCC interaction results in better outgrowth of both SCs and following axons. What about the inhibitor receptor Un5H? Like DCC, Unc5H reappears in adult SCS but after an injury these receptors, specifically the levels of subtypes Unc5H2 and Unc5H3, instead decline.

The netrin receptors, older "development" molecules, influence growth of an outgrowing SC "front," with axons to follow. After injury the attractive DCC receptors are tuned up, so that SCs can readily respond to netrin. The repulsive receptors, Unc5H2, are tuned down. What happens if you remove these receptors, for example using siRNA? When you "knockdown" DCC, growth of nerves is stunted. Both outgrowing SCs and following axons are unhappy. An attractive reason for growing, DCC, has now been shut off. In response, the inhibitory receptors Unc5H2 become more plentiful. Bad news abounds – the attractive receptor is gone and the repulsive receptors are congregating. What if you "knockdown" Unc5H2? If Unc5H2 is depleted by siRNA however, the situation begins to reverse itself. Outgrowth of both SCs and axons increases while DCC levels begin to increase. What transpires with DCC operates in the reverse direction to Unc5H2, in what is termed a "reciprocal" relationship. How these two receptors are so closely connected is not clear. We thus learn that there are two receptor molecules borrowed from early development, that influence growth of adult axons during regeneration. Their roles are similar to what they accomplish during development but they act in a strikingly different way in adults. Now they are located on SCs rather than neurons or axons. We are also reminded that SC support benefits axons whereas robust growing axons insist on the company of their SCs: a close and enduring partnership.

## Making new nerves: Summarizing the steps

Here we summarize the complexity and beauty of nerve regeneration (Fig. 36). Peripheral nerve wiring requires stability. However, some places in the body such as our skin, our gastrointestinal tract and bones have ongoing "wear and tear" over time. Thus, some element of ongoing growth is a requirement for nerves that service many body structures. This is likely not the case in the brain. New growth of brain axons might set up harmful circuits causing seizures or cognitive problems such as autism or confusion. Learning is an exception to this rule but its associated remodeling in the hippocampus is likely to be far more subtle than regenerative growth. The brain has the capacity to establish new patterns connecting existing

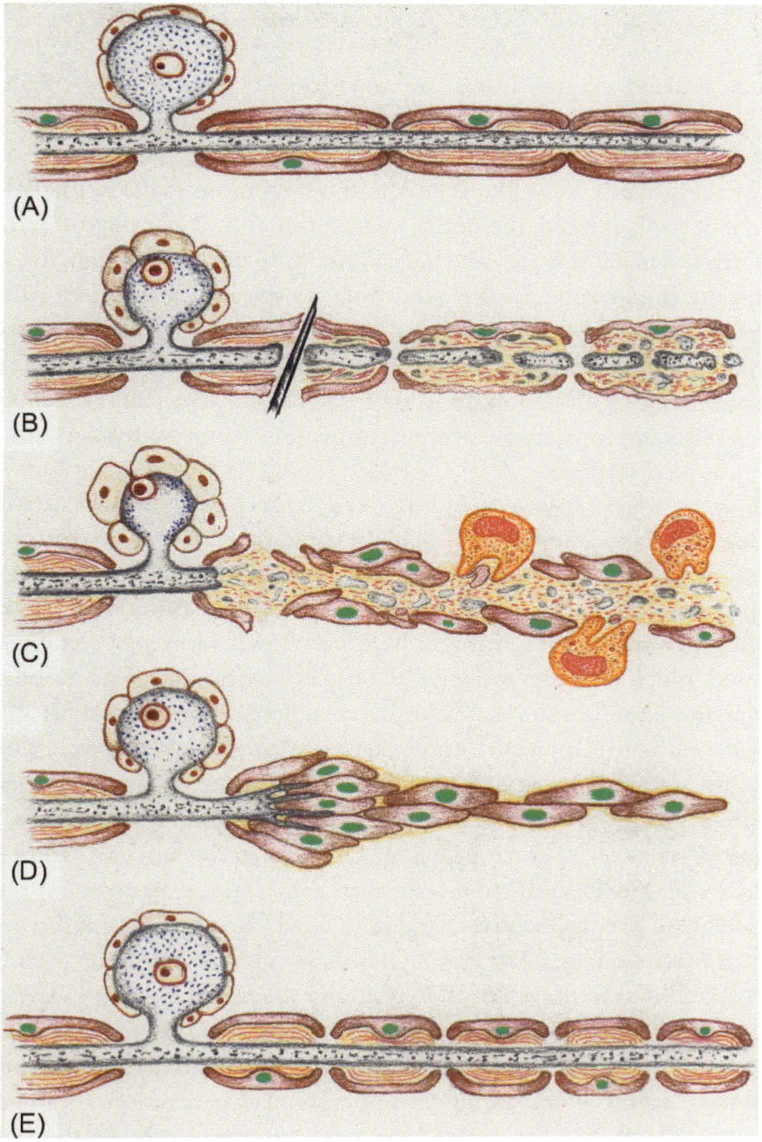


**Fig. 36** Illustration of sequential steps in the response of a sensory neuron to injury. (A) Normal intact nerve. (B) Early responses to nerve transection. These include movement of the neuron nucleus and neurofilaments (blue dots) to the side (cell body reaction), enlargement and proliferation of the perineuronal satellite glial cells, dissolution of the axons and their myelin in the distal stump. (C) Onset of sprouting from the proximal nerve stump, proliferation of Schwann cells and invasion of macrophages (brown with red nuclei). The cell body reaction and enlargement of perineuronal satellite glial cells are prominent. (D) Early sprouting. Multiple sprouts partnered with Schwann cells leave the proximal stump and enter the distal stump. (E) Maturation of successfully sprouted new axon with shortened myelin segments and recovery of the cell body response. Not shown is reconnection of the axon to its target tissue. Originally drawn by Scott Rogers and published in "Neurobiology of Peripheral Nerve Regeneration."[42]

or underutilized circuits instead of rewiring. To maintain the fidelity of their overall connections, peripheral neurons have retained a series of "anti-growth" molecules to suppress unwanted regeneration. Some operate at the level of the growth cone like RhoA and others more globally like PTEN. When a nerve is injured, the entire neuron tree from cell body to branches is rapidly aware of the fact and a calcium wave surges through the axon beyond the injury zone. In the axon distal to the injury an active destruction program is initiated involving a symphony of molecular participants. This is usually a favorable state of affairs, clearing the way for new axons and optimizing regeneration. In some instances, however, an AxD program is triggered inappropriately during neuropathies from neurotoxic drugs or autoimmune disorders.

Once an injury has occurred, the cell body, perikarya, alters its identity. Instead of a cell supporting stable wiring, it gears itself up for growth by activating RAGs. These regeneration associated genes synthesize specific proteins required for axon growth. During this time, "roadblock" or growth suppressing proteins might be tuned down. However, some remain unaltered and their very presence is obstructive. Several features of adult neurons thereby conspire to make nerve regeneration less robust than it could be. Growth suppressive molecules, loss of growth enhancing molecules, impaired interactions of growth cones with the basement membrane and other features of adulthood make them less plastic.

Several types of supporting cells are involved in regeneration. In ganglia, perineuronal satellite cells closely surround and support neurons. At the injury zone, SCs become activated and tune down their role in making myelin sheaths for stable axons. They proliferate and migrate out along a pathway that is established for new axons. In doing so they synthesize growth factors and lay down favorable basement membrane proteins that act as guidance routes for new axons. Activated SCs also autoengulf and degrade the myelin of the distal stump when axons are undergoing degeneration. Old residual myelin debris inhibits growth and must be removed. New myelin will be synthesized later as new regenerating axons mature.

Axons emerge from their injury zone with hesitation (Fig. 37). Once injured, the distal part of the axon cannot be simply reconnected to its parent proximal stump. Axonal degeneration ensues, SCs proliferate and migrate. New axon branches from the healthy proximal stump follow the SCs, emerge and enter the distal stump. These healthy axon sprouts, once instructed to grow, may send out many branches some of whom travel the wrong direction. For example some axons meant to connect to skin

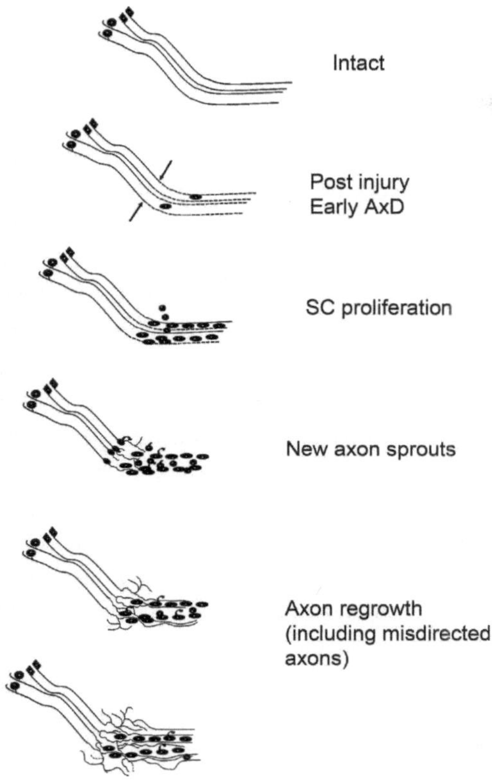


**Fig. 37** Scheme illustrating some early steps in regeneration following a nerve transection including early axonal degeneration, Schwann cell proliferation, formation of early sprouts then regrowth, including misdirected growth.

inadvertently travel into muscles. Some motor axons travel to the skin where they find no muscle fibers to connect with. These inappropriate branches may simply wither with time but how that happens is unclear. In embryos or neonates, extra unneeded branches are thought to degenerate and disappear because there are not enough growth factors available to keep them healthy. In adults this may not be the case since SCs remain readily available to provide support.

Once axons cross the injury zone, they meet with supportive SCs that have created pathways for them, the Bands of Bungner described earlier. New axons attempt to extend down to former targets in order to reconnect. The success of target reconnection has not been studied terribly well. Significant time is required to make new connections functional. It may be that new connections, while anatomically correct, never quite work as well as the old ones did. If it has been some time since the injury took place, the pathways for regeneration may not be very receptive to new axons and the muscle fibers may have atrophied. Muscle fibers with longstanding loss of their motor nerves, behave as if they are now uninterested in new axons. Sensory axons re-entering the skin may need to help reform specialized sensory organs like Merkel's discs. It may be that these important sensory organs also eventually disappear if axons are too long in coming.

If an axon is fortunate enough to reach its original target, some maturation is in order. Myelin must be reconstructed. All of the important ion channels at Nodes of Ranvier or on unmyelinated axons must be replaced. The correct placement, numbers and types are essential if sensory information is to be properly transmitted again. Branches in the skin require proper spacing. Motor axons need to accurately connect with abandoned muscle fibers in a 1:1 ratio so that any given muscle fiber receives only one branch.

An interesting study of nerve regeneration was conducted in humans after surgical repair of their accidentally severed hands.[180] Reconnecting hands is a major surgical feat and reconnecting nerves is one of several challenges. Investigators analyzed how well nerves had grown back into the skin five years after the surgical repair. Two of the three severed nerves of the hand underwent microscopic suture repair during the procedure. Unfortunately the long term innervation after periods of 26–81 months was limited. While the nerves had grown back to a degree, many areas of the skin had fewer than 50% of the original number of axons. Despite a successful surgical repair, axons failed to reach and establish connections with their targets during the prolonged time interval. It is highly likely that the rare but dramatic face transplants or the more common breast reconstructions after mastectomy have similar limitations. Successful and complete regeneration of peripheral axons remains a major clinical and research challenge.

The overall message is that there is a complex marshaling of events during regeneration involving axons, glial cells, inflammatory cells and others. The outcome is not guaranteed and besieged with complexities and pitfalls along the way. The success of regeneration has a major influence on the outcome of many neuropathies and is the key to neurological recovery. To achieve its potential, considerable work, expertise and investment in expertise is required. Repairing our wiring should no longer be "under the radar."

# CHAPTER 9

# Clinics and biology: Nerve heroes and the Peripheral Nerve Society

Here our story changes. A final pause with some contemplation will allow the reader to get an idea of how peripheral nerve specific research has emerged from the shadows. Many of the stories and discoveries highlighted arose from focus and fascination for structures that are unlike anything else in the nervous system, or in the body!

The structural building blocks and physiological attributes of nerves were discovered by a cohort of early biologists and neuroscientists, many mentioned above. Individual examples of foundational work by neuroscientists who sought to understand the unique biology of the PNS provide captivating accounts to those of us interested in the field. They include Ramon Y Cajal of Spain who described the events of nerve degeneration and regeneration.[181] Theodor Schwann described the Schwann cell. Jean Charcot and colleagues described inherited neuropathies known as CMTs (Charcot, Marie, Tooth disease). Guillain, Barré and Strohl described acute inflammatory demyelinating polyneuropathies, now referred to as the Guillain-Barré Syndrome. Tinel, known for the Tinel's sign, described nerve wounds in a famous monograph published in 1918.[182] Weir Mitchell, a neurologist who described nerve injures in the American Civil War is sometimes considered the true pioneer of peripheral nerve neurology.[98] There are many other examples, not described here.

By the 1970's, neurologists and neuroscientists began to realize that understanding the biology and diseases of the peripheral nervous system needed to be a full-time effort. There was no need to dilute their time in other areas of biology or neuroscience already well populated by investigators. Here we move to specifics around the birth of peripheral neurology that developed in the midst of other professional and scientific societies. For example, how did these fundamentals form the basis for a group of investigators exclusively devoted to nerves? Not heard of them? It is likely you have heard of the major American Neurological Societies, mainly comprised of physicians. Alternatively the Society for Neuroscience is a large and established

*Our Wired Nerves*
https://doi.org/10.1016/B978-0-12-821487-9.00009-X

international group largely populated by neuroscientists with lesser involvement by clinicians. In Canada we have the Canadian Neurological Sciences Federation and the Canadian Association of Neurosciences. Not intentionally and despite their size, there is a tendency for these groups to become siloed because of their consistent and large numbers of loyal members.

Quietly among these dominant groups and their annual meetings, a cluster of physicians and scientists decided to focus on peripheral nerve biology and disease. Originally called the "Peripheral Nerve Club," this eclectic group gradually brought many others into the fold including new and promising trainees who might carry the torch. A transplanted Canadian physician working at the Mayo Clinic in Rochester, Minnesota, Dr. Peter James Dyck (PJD for short) led the charge (Fig. 38). Beginning in the early 1960s Dr. Dyck investigated peripheral nerves at a number of levels. His contributions eventually became so extensive that he is also considered a "father" to the field. He started with reports on families with inherited neuropathies we have discussed called Charcot-Marie-Tooth disease (CMT), initially classified by Dyck as Hereditary Motor and Sensory Neuropathies (HMSNs). He and colleagues went on field trips to interview and examine patients and family members with these disorders and identified early patterns of their inheritance. These must have been interesting trips, some including portable nerve conduction equipment. His work progressed to provide early and definitive descriptions of autoimmune neuropathies including chronic inflammatory demyelinating polyneuropathy (CIDP). Peter Dyck combined careful and unique types of neurological examination with

**Fig. 38** Dr. Peter James Dyck, Mayo Clinic. *(With permission of Copyright holder, Dr. Peter James Dyck.)*

rigorous evaluation of nerve biopsies in patients. Overall these seminal investigations formed the basis for our pathological understanding and classification of neuropathies. At the same time in London UK, at the Royal Free Hospital and The National Hospital, Queen Square, another physician combined similar skills and equal rigor to the field, Dr. Peter K. Thomas (PK). PK and PJD were probably gentle competitors but also friends and colleagues and it is this chemistry that generated the "Peripheral Nerve Club." PK introduced the neurogeneticist Dr. Anita Harding to the field and her important contributions involved careful analysis of genetic neuropathies. Also around the same time, Dr. Arthur Asbury from University of Pennsylvania School of Medicine had made an early mark on the field publishing a nerve pathology text and writing seminal papers on inflammatory neuropathies and diabetic neuropathy complications. Dr. Asbury is considered a third seminal member of the Club.

As more peripheral nerve aficionados emerged, a small core committee of investigators took shape and called themselves the "Peripheral Nerve Study Group" (PNSG). Its first meeting was held in Carville, Louisiana in 1973 and by that time included Dr. Austin Sumner, an electrophysiologist often known for the "Lewis-Sumner" Syndrome, a type of inflammatory neuropathy and Dr. Gerard Said, a neurologist in Paris who had made a number of key descriptions of neuropathy. At its height in the late 1980s, PNSG meetings brought together groups of 100–120 participants in such places as Lake Couchiching Ontario Canada (my first meeting, organized by Dr. Tom Feasby) to Boppard Germany. Dr. John Griffin, a neurologist and seminal member from Johns Hopkins, described features of inflammatory neuropathy and axonal degeneration. The field continued to expand. Dr. Klaus Toyka, also a neurologist, set up a highly productive group of peripheral nerve investigators in Wurzburg, Germany that focused on inflammatory neuropathies. Dr. Michael Schröder, a neuropathologist from Aachen helped to define key pathological changes in nerves. Some of the important texts authored by this group included "Peripheral Neuropathy" edited by Dyck and Thomas (4 editions have been published since the first in 1975),[11] "Pathology of Peripheral Nerves" by Schröder,[183] "Pathology of Peripheral Nerves" by Asbury and Johnson,[184] "The Fine Structure of the Nervous System "by Peters, Webster and Palay.[185] In the mid 1980s, a competitor of the PNSG was built by Dr. Dyck and his Mayo Clinic colleagues Drs. Phillip Low and Anthony Windebank called the Peripheral Nerve Association of America (PNAA), later abbreviated to PNA (Peripheral Nerve Association). While PNSG tended to have a European and international research character, PNA and PNAA developed an educational platform. Many investigators had difficulty navigating the two groups and eventually convinced the organizations to merge. The Peripheral Nerve Society (PNS) [https://www.pnsociety.com/] was formed in 1993. PNS has grown in size and stature since. It is affiliated with the Journal of the Peripheral Nervous System, currently edited by Dr. Giuseppe Lauria from Milan. To give an idea of the extent of the growth, PNS in 2019, Genoa, hosted over a thousand neuroscientists, neurologists, neuropathologists and trainees incorporating research and education into an outstanding meeting and organization. This meeting, previously biannual, has now grown into a yearly meeting with an expanding and enthusiastic membership.

Parallel meetings and history have evolved for the specialty of peripheral nerve surgery not discussed here but that have included major contributors such as Drs. Sidney Sunderland,[2] Herbert Seddon, Göran Lundborg and others. A colleague, Dr. Rajiv Midha from the University of Calgary, has been a leader in these efforts.

I have a list of personal nerve heroes that are worth briefly naming here at the risk of missing many more. Some are mentioned already. Peter Dyck, continuing to work in his 10th decade, brought people together. Despite the early complexity of PNAA versus PNSG, PJD set up the foundations for PNS. He is a mentor to many of us. What I have appreciated have been his critical rigor, insistence on quantitation and his systematic and innovative approaches that have included classification of damaged axons, high quality nerve biopsy processing, and computerized quantitative sensory testing. I had less direct interactions with PK Thomas but recall him as a quiet but razor sharp clinician, with impressive critical thinking. Dr. Thomas Feasby, a former colleague, Department head and Dean, introduced me to Neurology and nerve disease. Tom is a collegial and supportive leader, quick to recognize nuances and build groups. He and Dr. William Brown, an electrophysiologist, conclusively demonstrated that weakness in Guillain-Barré syndrome patients is a direct result of conduction block in demyelinated nerves. He and Dr. Angelika Hahn led an impressive and collaborative research effort in the 1980s and 1990s that established the role of blood borne antibodies in the development of Guillain-Barré syndrome. Dr. Hahn, an outstanding, rigorous, and kind mentor no doubt inspired me to pursue this field. Her clinical and pathological skills are unmatched. Dr. Jack Griffin from Johns Hopkins was beloved by his colleagues for his intuitive insights and gentle manner. His observations around pathogenetic mechanisms of nerve disease have been foundational. Dr. David Cornblath from Johns Hopkins is an ambassador for the PNS, in large part responsible for its success and growth. Of his many qualities I appreciate are his critical honesty, support for new colleagues and contributions to many facets of neuropathy research including many original projects addressing inflammatory neuropathy. Dr. Tom Brushart from Johns Hopkins is the epitome of an academic surgeon, combining an orthopedic and nerve surgical practice with innovative and imaginative nerve science. His book "Nerve Repair" is a tour de force for all nerve surgeons and nerve biologists.[38]

I will only briefly mention further heroes of the peripheral nervous system that have been central to the field. Dr. Richard Hughes from London has been a driving force in operationalizing clinical trials of therapy for inflammatory neuropathies. Dr. Hugh Willison from Glasgow is been a leader in understanding autoimmune neuropathies. Drs. Patricia Armati and John Pollard have been an exceptional team from Sydney focusing on the biology of the Schwann cell and inflammatory neuropathies respectively. Dr. Jean-Michel Vallat from Limoges has made many contributions to peripheral neurobiology including its ultrastructural appreciation. Drs. Mary Reilly, Steven Scherer and Michael Shy have made major advances in the biology and understanding of inherited neuropathies. Dr. Richard Lewis, a codiscoverer of the Lewis-Sumner syndrome, has been a trailblazer in clinical trials of inflammatory neuropathies. Drs. Giuseppe Lauria from Milan and Michael Polydefkis from Johns Hopkins have been leaders in the use of diagnostic skin biopsies for analyzing epidermal innervation in neuropathy patients. Drs. Pieter Van Doorn, Bart Jacobs and colleagues from Rotterdam have spearheaded international efforts to understand GBS. Drs. Eva Feldman (University of Michigan), Paul Fernyhough (University of Manitoba), Roy Freeman (Harvard), Soroku Yagihashi (Hirosaki University), Rayaz Malik (Qatar) and colleagues have advanced the field of diabetic neuropathy including autonomic dysfunction in diabetes. Dr. Phillip Low from Mayo Clinic has been a leader the understanding and testing of autonomic neuropathies. Dr. Tessa Gordon formerly from the University of Alberta, is a neuroscientist who has led the field of motor axon plasticity. She and Tom Brushart worked on preferential motor reinnervation, described earlier in this book. Dr. Valerie Verge from the University of Saskatchewan, a longstanding colleague, is an inspired neuroscientist who has provided key work on neurotrophic factors and the response of neurons to injury.

Finally I conclude with a special mention of Dr. Charles Bolton, a colleague, mentor and supervisor who has been a mainstay of clinical peripheral neurology over many years (Fig. 39). His numerous contributions include

**Fig. 39** Dr. Charles F. Bolton, Western University, Mayo Clinic, Queen's University. *(With permission of Dr. Charles Bolton.)*

the seminal work I have described on "Bolton's neuropathy" or critical illness polyneuropathy. Charlie has been a leading educator and investigator in clinical neurophysiology having published with colleagues seminal books on Neuromuscular Diseases,[186] the Neurology of breathing,[187] Uremic neuropathy[188] and others. He has taught many how to safely perform needle electromyographic examinations of the diaphragm. Through his career at Western University in London, Canada, Mayo clinic and most recently at Queen's University, Kingston, Canada he has trained and inspired many.

A number of other clinicians and neuroscientists have made their contributions to peripheral nerve research. I cannot acknowledge all of them in this book but the field acknowledges them. Our work continues.

# Conclusions

I complete this short book on our nerve connectome having had immersion for much of my career in the field. This perspective does bring a realization that many of the questions and problems around peripheral neurons, neuropathies and regeneration will far outlive my immersion. It is a quest of great rewards with a small but highly dedicated cohort of neuroscientists and clinicians who have ignored obstacles and have pushed ahead. Most have been outstanding colleagues and have made major differences in the field. The Peripheral Nerve Society continues to expand internationally and new collaborations will yield new knowledge. New techniques will be required to probe peripheral neurons but they must be built on the hard won morphological and electrophysiological expertise that has created the field. Peripheral neurobiology has a difficult time winning a place in the overall attention by neurosciences however and only recently have the doors to publishing this work opened wider. We need to do better. This book is my attempt to connect the beauty and challenges of peripheral nerves, their neurobiology to the public at large. Remember that without peripheral nerves, there is no movement and no sensation!

# References


1. Belle M, Godefroy D, Couly G, et al. Tridimensional visualization and analysis of early human development. *Cell*. 2017;169:161–173.
2. Sunderland S. *Nerves and Nerve Injuries*. 2nd ed. Edinburgh: Churchill Livingstone; 1978.
3. Court FA, Hendriks WT, MacGillavry HD, Alvarez J, van Minnen J. Schwann cell to axon transfer of ribosomes: toward a novel understanding of the role of gila in the nervous system. *J Neurosci*. 2008;28:11024–11029.
4. Waxman SG, Kocsis JD, Stys PK. *The Axon: Structure, Function and Pathophysiology*. Oxford: Oxford University Press; 1995.
5. Armati PJ. *The Biology of Schwann Cells*. Cambridge: Cambridge; 2007.
6. Rose PK, Cushing S. Relationship between morphoelectrotonic properties of motoneuron dendrites and their trajectory. *J Comp Neurol*. 2004;473:562–581.
7. Gesslbauer B, Hruby LA, Roche AD, et al. Axonal components of nerves innervating the human arm. *Ann Neurol*. 2017;82:396–408.
8. Barnett HJM, Foster JB, Hudgson P. *Syringomyelia*. Toronto: W.B. Saunders; 1973.
9. Campenot RB. *Animal Electricity*. Cambridge, MA: Harvard University Press; 2016.
10. Kandel ER, Schwartz JH, Jessell TM, et al. *Principles of Neural Science*. 5th ed. New York: McGraw-Hill Education; 2012.
11. Dyck PJ, Thomas PK. *Peripheral Neuropathy*. Philadelphia: Elsevier Saunders; 2005.
12. Barr ML, Kiernan JA. *The Human Nervous System*. 4th ed. Philadelphia: Harper and Row; 1983.
13. Scherer SS, Arroyo EJ. Recent progress on the molecular organization of myelinated axons. *J Peripher Nerv Syst*. 2002;7:1–12.
14. Fehmi J, Scherer SS, Willison HJ, et al. Nodes, paranodes and neuropathies. *J Neurol Neurosurg Psychiatry*. 2018;89:61–71.
15. Zylka MJ, Rice FL, Anderson DJ. Topographically distinct epidermal nociceptive circuits revealed by axonal tracers targeted to Mrgprd. *Neuron*. 2005;45:17–25.
16. Munger BL, Ide C. The structure and function of cutaneous sensory receptors. *Arch Histol Cytol*. 1988;51:1–34.
17. Lyman M. *The Remarkable Life of the Skin*. London: Bantam Press; 2019.
18. Low PA, Benarroch EE. *Clinical Autonomic Disorders*. Philadelphia: Wolters-Kluwer, Lippincott; 2008.

18a. Abdo H, Calvo-Enrique L, Lopez JM, Song J, Zhang MD, Usoskin D, El Manira A, Adameyko I, Hjerling-Leffler J, Ernfors P. Specialized cutaneous Schwann cells initiate pain sensation. *Science*. 2019;365(6454):695-699. https://doi.org/10.1126/science.aax6452. PMID: 31416963.

19. Willis DE, Twiss JL. The evolving roles of axonally synthesized proteins in regeneration. *Curr Opin Neurobiol*. 2006;16:111–118.
20. Piper M, Holt C. RNA translation in axons. *Annu Rev Cell Dev Biol*. 2004;20:505–523.
21. Gordon IT, Whelan PJ. Deciphering the organization and modulation of spinal locomotor central pattern generators. *J Exp Biol*. 2006;209:2007–2014.
22. Gordon IT, Whelan PJ. Monoaminergic control of cauda-equina-evoked locomotion in the neonatal mouse spinal cord. *J Neurophysiol*. 2006;96:3122–3129.
23. Abraira VE, Ginty DD. The sensory neurons of touch. *Neuron*. 2013;79:618–639.
24. Johnson KO, Lamb GD. Neural mechanisms of spatial tactile discrimination: neural patterns evoked by braille-like dot patterns in the monkey. *J Physiol*. 1981;310:117–144.
25. Bolton CF, Winkelmann RK, Dyck PJ. A quantitative study of Meissner's corpuscles in man. *Neurology*. 1966;16:1–9.

26. Spencer PS, Schaumburg HH. An ultrastructural study of the inner core of the Pacinian corpuscle. *J Neurocytol.* 1973;2:217–235.
27. Vrontou S, Wong AM, Rau KK, et al. Genetic identification of C fibres that detect massage-like stroking of hairy skin in vivo. *Nature.* 2013;493:669–673.
28. Wang L, Zhou H, Zhang M, et al. Structure and mechanogating of the mammalian tactile channel PIEZO2. *Nature.* 2019;573:225–229.
29. Lin YC, Guo YR, Miyagi A, et al. Force-induced conformational changes in PIEZO1. *Nature.* 2019;573:230–234.
30. Chesler AT, Szczot M, Bharucha-Goebel D, et al. The role of PIEZO2 in human mechanosensation. *N Engl J Med.* 2016;375:1355–1364.
31. Roper SD, Chaudhari N. Taste buds: cells, signals and synapses. *Nat Rev Neurosci.* 2017;18:485–497.
32. Koizumi A, Tsuchiya A, Nakajima K, et al. Human sweet taste receptor mediates acid-induced sweetness of miraculin. *Proc Natl Acad Sci U S A.* 2011;108:16819–16824.
33. Degrace-Passilly P, Besnard P. CD36 and taste of fat. *Curr Opin Clin Nutr Metab Care.* 2012;15:107–111.
34. Mishra SK, Hoon MA. The cells and circuitry for itch responses in mice. *Science.* 2013;340:968–971.
35. Gerdts J, Summers DW, Milbrandt J, et al. Axon self-destruction: new links among SARM1, MAPKs, and $NAD^+$ metabolism. *Neuron.* 2016;89:449–460.
36. Woo V, Cheng C, Duraikannu A, et al. Caspase-6 is a dispensable enabler of adult mammalian axonal degeneration. *Neuroscience.* 2018;371:242–253.
37. Witzel C, Rohde C, Brushart TM. Pathway sampling by regenerating peripheral axons. *J Comp Neurol.* 2005;485:183–190.
38. Brushart TM. *Nerve Repair.* Oxford: Oxford University Press; 2011.
39. Stoll G, Wilder-Smith E, Bendszus M. Imaging of the peripheral nervous system. *Handb Clin Neurol.* 2013;115:137–153.
40. O'Brien M. *Aids to the Examination of the Peripheral Nervous System.* 4th ed. London: Saunders Elsevier; 2010.
41. Theriault M, Dort J, Sutherland G, et al. A prospective quantitative study of sensory deficits after whole sural nerve biopsies in diabetic and nondiabetic patients. Surgical approach and the role of collateral sprouting. *Neurology.* 1998;50:480–484.
42. Zochodne DW. *Neurobiology of Peripheral Nerve Regeneration.* Cambridge: Cambridge University Press; 2008.
43. Onishi A, O'Brien PC, Dyck PJ. Studies to improve fixation of human nerves. 2. Effect of time elapsed between death and glutaraldehyde fixation on relationship of axonal area to number of myelin lamellae. *J Neurol Sci.* 1974;23:387–390.
44. Stewart JD. *Focal Peripheral Neuropathies.* 4th ed. West Vancouver: JBJ Publishing; 2010.
45. Zochodne DW, Semmler RT, Ludwin SK, et al. Acute fulminant symmetrical vasculitic polyneuropathy: need for early biopsy. *Clin Neuropathol.* 1996;15:113–115.
46. Yang W, Lu J, Weng J, et al. Prevalence of diabetes among men and women in China. *N Engl J Med.* 2010;362:1090–1101.
47. Shapiro AM, Ricordi C, Hering BJ, et al. International trial of the Edmonton protocol for islet transplantation. *N Engl J Med.* 2006;355:1318–1330.
48. Schaumburg HH, Wisniewski HM, Spencer PS. Ultrastructural studies of the dying-back process. I. Peripheral nerve terminal and axon degeneration in systemic acrylamide intoxication. *J Neuropathol Exp Neurol.* 1974;33:260–284.
49. Fried K, Govrin-Lippmann R, Rosenthal F, et al. Ultrastructure of afferent axon endings in a neuroma. *J Neurocytol.* 1991;20:682–701.
50. Devor M, Govrin-Lippmann R, Angelides K. $Na^+$ channel immunolocalization in peripheral mammalian axons and changes following nerve injury and neuroma formation. *J Neurosci.* 1993;13:1976–1992.

51. Gordon T, Hegedus J, Tam SL. Adaptive and maladaptive motor axonal sprouting in aging and motoneuron disease. *Neurol Res*. 2004;26:174–185.
52. Asbury AK, Aldredge H, Hershberg R, et al. Oculomotor palsy in diabetes mellitus: a clinico-pathological study. *Brain*. 1970;93:555–566.
53. Stewart JD. Diabetic truncal neuropathy: topography of the sensory deficit. *Ann Neurol*. 1989;25:233–238.
54. Dyck PJ, Norell JE, Dyck PJ. Microvasculitis and ischemia in diabetic lumbosacral radiculoplexus neuropathy. *Neurology*. 1999;53:2113–2121.
55. Raff MC, Sangalang V, Asbury AK. Ischemic mononeuropathy multiplex associated with diabetes mellitus. *Arch Neurol*. 1968;18:487–499.
56. Chalk C, Benstead TJ, Moore F. Aldose reductase inhibitors for the treatment of diabetic polyneuropathy. *Cochrane Database Syst Rev*. 2007. CD004572.
57. Brownlee M, Cerami A, Vlassara H. Advanced glycosylation end products in tissue and the biochemical basis of diabetic. *N Engl J Med*. 1988;318:1315–1321.
58. Feldman EL, Callaghan BC, Pop-Busui R, et al. Diabetic neuropathy. *Nat Rev Dis Primers*. 2019;5:41.
59. Frazier WA, Angeletti RH, Bradshaw RA. Nerve growth factor and insulin. *Science*. 1972;176:482–488.
60. Zochodne DW. Sensory neurodegeneration in diabetes: beyond glucotoxicity. *Int Rev Neurobiol*. 2016;127:151–180.
61. Singh B, Xu Y, McLaughlin T, et al. Resistance to trophic neurite outgrowth of sensory neurons exposed to insulin. *J Neurochem*. 2012;121:253–276.
62. Cheng C, Kobayashi M, Martinez JA, et al. Evidence for epigenetic regulation of gene expression and function in chronic experimental diabetic neuropathy. *J Neuropathol Exp Neurol*. 2015;74:804–817.
63. de Freitas MR, Said G. Leprous neuropathy. *Handb Clin Neurol*. 2013;115:499–514.
64. Sasakawa Y. *No Matter Where the Journey Takes Me*. London: Hurst and Company; 2019.
65. Fischer M. Leprosy—an overview of clinical features, diagnosis, and treatment. *J Dtsch Dermatol Ges*. 2017;15:801–827.
66. Antia NH, Shetty VP. *The Peripheral Nerve in Leprosy and Other Neuropathies*. Calcutta: Oxford University Press; 1997.
67. Nascimento OJ. Leprosy neuropathy: clinical presentations. *Arq Neuropsiquiatr*. 2013;71:661–666.
68. van Brakel WH, Khawas IB. Silent neuropathy in leprosy: an epidemiological description. *Lepr Rev*. 1994;65:350–360.
69. Jacobs BC, van den Berg B, Verboon C, et al. International Guillain-Barré Syndrome Outcome Study: protocol of a prospective observational cohort study on clinical and biological predictors of disease course and outcome in Guillain-Barré syndrome. *J Peripher Nerv Syst*. 2017;22:68–76.
70. Feasby TE, Gilbert JJ, Brown WF, et al. An acute axonal form of Guillain-Barré polyneuropathy. *Brain*. 1986;109:1115–1126.
71. McKhann GM, Cornblath DR, Griffin JW, et al. Acute motor axonal neuropathy: a frequent cause of acute flaccid paralysis in China. *Ann Neurol*. 1993;33(4):333–342.
72. Langton BS, Ondrich S, Hill P. *Guillain-Barré Syndrome: 5 Years later*. Victoria, BC: Trafford Publishing; 2006.
73. Langton BS. *A First Step: Understanding Guillain-Barré Syndrome*. Victoria, BC: Trafford Publishing; 2002.
74. Dyck PJ, Lais AC, Ohta M, et al. Chronic inflammatory polyradiculoneuropathy. *Mayo Clin Proc*. 1975;50:621–637.
75. Pascual-Goni E, Martin-Aguilar L, Querol L. Autoantibodies in chronic inflammatory demyelinating polyradiculoneuropathy. *Curr Opin Neurol*. 2019;32:651–657.
76. Manso C, Querol L, Lleixa C, et al. Anti-Neurofascin-155 IgG4 antibodies prevent paranodal complex formation in vivo. *J Clin Invest*. 2019;129:2222–2236.

77. Manso C, Querol L, Mekaouche M, et al. Contactin-1 IgG4 antibodies cause paranode dismantling and conduction defects. *Brain*. 2016;139:1700–1712.
78. Bolton CF, Gilbert JJ, Hahn AF, et al. Polyneuropathy in critically ill patients. *J Neurol Neurosurg Psychiatry*. 1984;47:1223–1231.
79. Zochodne DW, Bolton CF, Wells GA, et al. Critical illness polyneuropathy a complication of sepsis and multiple organ failure. *Brain*. 1987;110:819–841.
80. Staff NP, Grisold A, Grisold W, et al. Chemotherapy-induced peripheral neuropathy: A current review. *Ann Neurol*. 2017;81:772–781.
81. Windebank AJ, Grisold W. Chemotherapy-induced neuropathy. *J Peripher Nerv Syst*. 2008;13:27–46.
82. Anderson D, Beecher G, Nathoo N, et al. Proposed diagnostic and treatment paradigm for high-grade neurological complications of immune checkpoint inhibitors. *Neurooncol Pract*. 2019;6:340–345.
83. Jarand J, Zochodne DW, Martin LO, et al. Neurological complications of infliximab. *J Rheumatol*. 2006;33:1018–1020.
84. Pardo CA, McArthur JC, Griffin JW. HIV neuropathy: insights in the pathology of HIV peripheral nerve disease. *J Peripher Nerv Syst*. 2001;6:21–27.
85. Andrade C. A peculiar form of peripheral neuropathy. Familiar atypical generalized amyloidosis with special involvement of the peripheral nerves. *Brain*. 1952;75:408–426.
86. Benson MD, Kincaid JC. The molecular biology and clinical features of amyloid neuropathy. *Muscle Nerve*. 2007;36:411–423.
87. Quarta CC, Solomon SD. Stabilizing transthyretin to treat ATTR cardiomyopathy. *N Engl J Med*. 2018;379:1083–1084.
88. Benson MD, Waddington-Cruz M, Berk JL, et al. Inotersen treatment for patients with hereditary transthyretin amyloidosis. *N Engl J Med*. 2018;379:22–31.
89. Adams D, Hawkins PN, Polydefkis M. Oligonucleotide drugs for transthyretin amyloidosis. *N Engl J Med*. 2018;379:2086.
90. Pipis M, Rossor AM, Laura M, et al. Next-generation sequencing in Charcot-Marie-Tooth disease: opportunities and challenges. *Nat Rev Neurol*. 2019;15:644–656.
91. Laura M, Pipis M, Rossor AM, et al. Charcot-Marie-tooth disease and related disorders: an evolving landscape. *Curr Opin Neurol*. 2019;32:641–650.
92. Mathis S, Vallat JM. Jules Dejerine and the peripheral nervous system. *Neurology*. 2017;89:611–615.
93. Dyck PJ, Lambert EH. Lower motor and primary sensory neuron diseases with peroneal muscular atrophy. I. Neurologic, genetic, and electrophysiologic findings in hereditary polyneuropathies. *Arch Neurol*. 1968;18:603–618.
94. Sacks O. *The Man Who Mistook His Wife for a Hat*. New York: Simon and Schuster; 1985.
95. Eystathioy T, Chan EK, Takeuchi K, et al. Clinical and serological associations of autoantibodies to GW bodies and a novel cytoplasmic autoantigen GW182. *J Mol Med*. 2003;81:811–818.
96. Li S, Lian SL, Moser JJ, et al. Identification of GW182 and its novel isoform TNGW1 as translational repressors in Ago2-mediated silencing. *J Cell Sci*. 2008;121:4134–4144.
97. Treede RD, Jensen TS, Campbell JN, et al. Neuropathic pain: redefinition and a grading system for clinical and research purposes. *Neurology*. 2008;70:1630–1635.
98. Mitchell SW, Morehouse GR, Keen WW. *Gunshot Wounds and Other Injuries of Nerves*. 1864 (original publication date). San Francisco: Norman Publishing; 1989.
99. Apfel SC, Kessler JA, Adornato BT, et al. Recombinant human nerve growth factor in the treatment of diabetic polyneuropathy. NGF Study Group. *Neurology*. 1998;51:695-702.
100. Asbury AK, Fields HL. Pain due to peripheral nerve damage: an hypothesis. *Neurology*. 1984;34:1587–1590.
101. Bove GM, Light AR. The nervi nervorum, missing link for neuropathic pain? *Pain Forum*. 1997;6:181–190.

102. Zochodne DW. Epineurial peptides: a role in neuropathic pain? *Can J Neurol Sci.* 1993;20:69–72.
103. Devor M. Neuropathic pain and injured nerve: peripheral mechanisms. *Br Med Bull.* 1991;47:619–630.
104. Ochoa JL. Essence, investigation, and management of "neuropathic" pains: hopes from acknowledgment of chaos. *Muscle Nerve.* 1993;16:997–1008.
105. Cline MA, Ochoa J, Torebjork HE. Chronic hyperalgesia and skin warming caused by sensitized C nociceptors. *Brain.* 1989;112(Pt 3):621–647.
106. Lax H, Zochodne DW. Causalgic median mononeuropathies: segmental rubror and edema. *Muscle Nerve.* 1995;18:245–247.
107. Schwab ME, Strittmatter SM. Nogo limits neural plasticity and recovery from injury. *Curr Opin Neurobiol.* 2014;27:53–60.
108. Kater S, Letourneau P, eds. *Biology of the Nerve Growth Cone.* New York: Alan R. Liss; 1985.
109. Letourneau PC, Kater SB, Macagno ER. *The Nerve Growth Cone.* New York: Raven Press; 1991.
110. Gordon-Weeks P. *Neuronal Growth Cones.* Cambridge: Cambridge University Press; 2000.
111. Langley JN. Note on regeneration of prae-ganglionic fibres of the sympathetic. *J Physiol.* 1895;18:280–284.
112. Sperry RW. Chemoaffinity in the orderly growth of nerve fiber patterns and connections. *Proc Natl Acad Sci U S A.* 1963;50:703–710.
113. Zipursky SL, Sanes JR. Chemoaffinity revisited: dscams, protocadherins, and neural circuit assembly. *Cell.* 2010;143:343–353.
114. Brushart TM. Preferential reinnervation of motor nerves by regenerating motor axons. *J Neurosci.* 1988;8:1026–1031.
115. Brushart TM. Motor axons preferentially reinnervate motor pathways. *J Neurosci.* 1993;13:2730–2738.
116. Brushart TM, Jari R, Verge V, et al. Electrical stimulation restores the specificity of sensory axon regeneration. *Exp Neurol.* 2005;194:221–229.
117. Saito H, Nakao Y, Takayama S, et al. Specific expression of an HNK-1 carbohydrate epitope and NCAM on femoral nerve Schwann cells in mice. *Neurosci Res.* 2005;53:314–322.
118. Hoke A, Redett R, Hameed H, et al. Schwann cells express motor and sensory phenotypes that regulate axon regeneration. *J Neurosci.* 2006;26:9646–9655.
119. Franz CK, Rutishauser U, Rafuse VF. Intrinsic neuronal properties control selective targeting of regenerating motoneurons. *Brain.* 2008;131:1492–1505.
120. Al-Majed AA, Neumann CM, Brushart TM, et al. Brief electrical stimulation promotes the speed and accuracy of motor axonal regeneration. *J Neurosci.* 2000;20:2602–2608.
121. Chen YY, McDonald D, Cheng C, et al. Axon and Schwann cell partnership during nerve regrowth. *J Neuropathol Exp Neurol.* 2005;64:613–622.
122. McDonald D, Cheng C, Chen Y, et al. Early events of peripheral nerve regeneration. *Neuron Glia Biol.* 2006;2:139–147.
123. Morris JH, Hudson AR, Weddell G. A study of degeneration and regeneration in the divided rat sciatic nerve based on electron microscopy. IV. Changes in fascicular microtopography, perineurium and endoneurial fibroblasts. *Z Zellforsch Mikrosk Anat.* 1972;124:165–203.
124. Morris JH, Hudson AR, Weddell G. A study of degeneration and regeneration in the divided rat sciatic nerve based on electron microscopy. I. The traumatic degeneration of myelin in the proximal stump of the divided nerve. *Z Zellforsch Mikrosk Anat.* 1972;124:76–102.
125. Morris JH, Hudson AR, Weddell G. A study of degeneration and regeneration in the divided rat sciatic nerve based on electron microscopy. II. The development of the "regenerating unit". *Z Zellforsch Mikrosk Anat.* 1972;124:103–130.

126. Morris JH, Hudson AR, Weddell G. A study of degeneration and regeneration in the divided rat sciatic nerve based on electron microscopy. III. Changes in the axons of the proximal stump. *Z Zellforsch Mikrosk Anat.* 1972;124:131–164.
127. Gardiner NJ. Integrins and the extracellular matrix: key mediators of development and regeneration of the sensory nervous system. *Dev Neurobiol.* 2011;71:1054–1072.
128. Lemons ML, Condic ML. Integrin signaling is integral to regeneration. *Exp Neurol.* 2008;209:343–352.
129. Liu W-Q, Martinez JA, Durand J, et al. RGD-mediated adhesive interactions are important for peripheral axon outgrowth in vivo. *Neurobiol Dis.* 2009;34:11–22.
130. Mei L, Nave KA. Neuregulin-ERBB signaling in the nervous system and neuropsychiatric diseases. *Neuron.* 2014;83:27–49.
131. Birchmeier C, Bennett DL. Neuregulin/ErbB signaling in developmental myelin formation and nerve repair. *Curr Top Dev Biol.* 2016;116:45–64.
132. Stassart RM, Fledrich R, Velanac V, et al. A role for Schwann cell-derived neuregulin-1 in remyelination. *Nat Neurosci.* 2013;16:48–54.
133. Zochodne DW, Ho LT. Influence of perivascular peptides on endoneurial blood flow and microvascular resistance in the sciatic nerve of the rat. *J Physiol.* 1991;444:615–630.
134. Toth C, Willis D, Twiss JL, et al. Locally synthesized calcitonin gene-related peptide has a critical role in peripheral nerve regeneration. *J Neuropathol Exp Neurol.* 2009;68:326–337.
135. Cheng C, Webber CA, Wang J, et al. Activated RHOA and peripheral axon regeneration. *Exp Neurol.* 2008;212:358–369.
136. Levi-Montalcini R, Hamburger V. Selective growth stimulating effects of mouse sarcoma on the sensory and sympathetic nervous system of the chick embryo. *J Exp Zool.* 1951;116:321–362.
137. Levi-Montalcini R, Calissano P. The nerve-growth factor. *Sci Am.* 1979;240:68–77.
138. Cheng C, Guo GF, Martinez JA, et al. Dynamic plasticity of axons within a cutaneous milieu. *J Neurosci.* 2010;30:14735–14744.
139. Fernyhough P, Willars GB, Lindsay RM, et al. Insulin and insulin-like growth factor I enhance regeneration in cultured adult rat sensory neurones. *Brain Res.* 1993;607:117–124.
140. Xu QG, Li X-Q, Kotecha SA, et al. Insulin as an in vivo growth factor. *Exp Neurol.* 2004;188:43–51.
141. Guo G, Kan M, Martinez JA, et al. Local insulin and the rapid regrowth of diabetic epidermal axons. *Neurobiol Dis.* 2011;43:414–421.
142. Singhal A, Cheng C, Sun H, et al. Near nerve local insulin prevents conduction slowing in experimental diabetes. *Brain Res.* 1997;763:209–214.
143. Brussee V, Cunningham FA, Zochodne DW. Direct insulin signaling of neurons reverses diabetic neuropathy. *Diabetes.* 2004;53:1824–1830.
144. Krishnan A, Bhavanam S, Zochodne D. An intimate role for adult dorsal root ganglia resident cycling cells in the generation of local macrophages and satellite glial cells. *J Neuropathol Exp Neurol.* 2018;77:929–941.
145. Zochodne DW, Nguyen C, Sharkey KA. Accumulation and degranulation of mast cells in experimental neuromas. *Neurosci Lett.* 1994;182:3–6.
146. Olsson Y. Degranulation of mast cells in peripheral nerve injuries. *Acta Neurol Scand.* 1967;43:365–374.
147. Lindborg JA, Mack M, Zigmond RE. Neutrophils are critical for myelin removal in a peripheral nerve injury model of wallerian degeneration. *J Neurosci.* 2017;37:10258–10277.
148. Fry EJ, Ho C, David S. A role for Nogo receptor in macrophage clearance from injured peripheral nerve. *Neuron.* 2007;53:649–662.
149. Murphy PG, Grondin J, Altares M, et al. Induction of interleukin-6 in axotomized sensory neurons. *J Neurosci.* 1995;15:5130–5138.

150. McDonald DS, Cheng C, Martinez JA, et al. Regenerative arrest of inflamed peripheral nerves: role of nitric oxide. *Neuroreport*. 2007;18:1635–1640.
151. Perlson E, Medzihradszky KF, Darula Z, et al. Differential proteomics reveals multiple components in retrogradely transported axoplasm after nerve injury. *Mol Cell Proteomics*. 2004;3:510–520.
152. Lieberman AR. The axon reaction: a review of the principal features of perikaryal responses to axon injury. *Int Rev Neurobiol*. 1971;14:49–124.
153. Verge VM, Gratto KA, Karchewski LA, et al. Neurotrophins and nerve injury in the adult. *Philos Trans R Soc Lond B Biol Sci*. 1996;351:423–430.
154. Coggins PJ, Zwiers H. B-50 (GAP-43): biochemistry and functional neurochemistry of a neuron-specific phosphoprotein. *J Neurochem*. 1991;56:1095–1106.
155. Skene JHP. Axonal growth-associated proteins. *Annu Rev Neurosci*. 1989;12:127–156.
156. Geremia NM, Gordon T, Brushart TM, et al. Electrical stimulation promotes sensory neuron regeneration and growth-associated gene expression. *Exp Neurol*. 2007;205:347–359.
157. Singh B, Xu QG, Franz CK, et al. Accelerated axon outgrowth, guidance, and target reinnervation across nerve transection gaps following a brief electrical stimulation paradigm. *J Neurosurg*. 2012;116:498–512.
158. Gordon T, Chan KM, Sulaiman OA, et al. Accelerating axon growth to overcome limitations in functional recovery after peripheral nerve injury. *Neurosurgery*. 2009;65:A132–A144.
159. Wong JN, Olson JL, Morhart MJ, et al. Electrical stimulation enhances sensory recovery: a randomized controlled trial. *Ann Neurol*. 2015;77:996–1006.
160. Kuhn T. *The Structure of Scientific Revolutions*. Chicago: University of Chicago Press; 1962.
161. Reynolds BA, Weiss S. Generation of neurons and astrocytes from isolated cells of the adult mammalian central nervous system. *Science*. 1992;255:1707–1710.
162. Sorrells SF, Paredes MF, Cebrian-Silla A, et al. Human hippocampal neurogenesis drops sharply in children to undetectable levels in adults. *Nature*. 2018;555:377–381.
163. Jakymiw A, Lian S, Eystathioy T, et al. Disruption of GW bodies impairs mammalian RNA interference. *Nat Cell Biol*. 2005;7:1267–1274.
164. Devor M, Govrin-Lippmann R, Frank I, et al. Proliferation of primary sensory neurons in adult rat dorsal root ganglion and the kinetics of retrograde cell loss after sciatic nerve section. *Somatosens Res*. 1985;3:139–167.
165. Toma JG, McKenzie IA, Bagli D, et al. Isolation and characterization of multipotent skin-derived precursors from human skin. *Stem Cells*. 2005;23:727–737.
166. McKenzie IA, Biernaskie J, Toma JG, et al. Skin-derived precursors generate myelinating Schwann cells for the injured and dysmyelinated nervous system. *J Neurosci*. 2006;26:6651–6660.
167. Khuong HT, Kumar R, Senjaya F, et al. Skin derived precursor Schwann cells improve behavioral recovery for acute and delayed nerve repair. *Exp Neurol*. 2014;254:168–179.
168. Shakhbazau A, Mohanty C, Kumar R, et al. Sensory recovery after cell therapy in peripheral nerve repair: effects of naive and skin precursor-derived Schwann cells. *J Neurosurg*. 2014;121:423–431.
169. Schmucker D, Chen B. Dscam and DSCAM: complex genes in simple animals, complex animals yet simple genes. *Genes Dev*. 2009;23:147–156.
170. Diamond J, Holmes M, Coughlin M. Endogenous NGF and nerve impulses regulate the collateral sprouting of sensory axons in the skin of the adult rat. *J Neurosci*. 1992;12:1454–1466.
171. Park KK, Liu K, Hu Y, et al. Promoting axon regeneration in the adult CNS by modulation of the PTEN/mTOR pathway. *Science*. 2008;322:963–966.
172. Christie KJ, Webber CA, Martinez JA, et al. PTEN inhibition to facilitate intrinsic regenerative outgrowth of adult peripheral axons. *J Neurosci*. 2010;30:9306–9315.

173. Tucker BA, Rahimtula M, Mearow KM. Laminin and growth factor receptor activation stimulates differential growth responses in subpopulations of adult DRG neurons. *Eur J Neurosci*. 2006;24:676–690.
174. Singh B, Singh V, Krishnan A, et al. Regeneration of diabetic axons is enhanced by selective knockdown of the PTEN gene. *Brain*. 2014;137:1051–1067.
175. Christie KJ, Krishnan A, Martinez JA, et al. Enhancing adult nerve regeneration through the knockdown of retinoblastoma protein. *Nat Commun*. 2014;5:3670.
176. Duraikannu A, Martinez JA, Chandrasekhar A, et al. Expression and manipulation of the APC-beta-catenin pathway during peripheral neuron regeneration. *Sci Rep*. 2018;8:13197.
177. Condic ML. Adult neuronal regeneration induced by transgenic integrin expression. *J Neurosci*. 2001;21:4782–4788.
178. Martinez JA, Kobayashi M, Krishnan A, et al. Intrinsic facilitation of adult peripheral nerve regeneration by the Sonic hedgehog morphogen. *Exp Neurol*. 2015;271:493–505.
179. Webber CA, Christie KJ, Cheng C, et al. Schwann cells direct peripheral nerve regeneration through the Netrin-1 receptors, DCC and Unc5H2. *Glia*. 2011;59:1503–1517.
180. Wiberg M, Hazari A, Ljungberg C, et al. Sensory recovery after hand reimplantation: a clinical, morphological, and neurophysiological study in humans. *Scand J Plast Reconstr Surg Hand Surg*. 2003;37:163–173.
181. Ramon y Cajal S. Degeneration and regeneration of the nervous system. In: DeFelipe J, Jones EG, eds. *Cajal's degeneration and regeneration of the nervous system*. Oxford: Oxford University Press; 1928.
182. Tinel J. *Nerve Wounds, Symptomatology of Peripheral Nerve Lesions Caused by War Wounds*. New York: William Wood and Company; 1918.
183. Schröder JM. *Pathology of Peripheral Nerves*. Berlin: Springer; 2001.
184. Asbury AK, Johnson PC. *Pathology of Peripheral Nerve*. Philadelphia: Saunders; 1978.
185. Peters A, Palay SL, Webster HD. *The Fine Structure of the Nervous System*. 3rd ed. New York: Oxford University Press; 1991.
186. Brown WF, Bolton CF, Aminoff MJ. *Neuromuscular Function and Disease*. Philadelphia: Elsevier Saunders; 2002.
187. Bolton CF, Chen R, Wijdicks EFM, et al. *Neurology of Breathing*. Toronto: Butterworth-Heinemann; 2003.
188. Bolton CF, Young GB. *Neurological Complications of Renal Disease*. Boston: Butterwortths; 1990.

# Index

Note: Page numbers followed by *f* indicate figures.

## H



## I



## J



## K



## L

## O



## P

FSC
www.fsc.org
MIX
Papier | Fördert gute Waldnutzung
FSC® C083411

Kolibri 360 GmbH
Ferdinand-Jühlke-Straße 7
99095 Erfurt, Deutschland
produktsicherheit@kolibri360.de